The Great American
Steamboat Race

The Great American Steamboat Race

The *Natchez* and the *Robert E. Lee* and the Climax of an Era

BENTON RAIN PATTERSON

McFarland & Company, Inc., Publishers
Jefferson, North Carolina, and London

LIBRARY OF CONGRESS CATALOGUING-IN-PUBLICATION DATA

Patterson, Benton Rain, 1929–
  The great American steamboat race : the Natchez and the Robert E. Lee and the climax of an era / Benton Rain Patterson.
    p.   cm.
  Includes bibliographical references and index.

  ISBN 978-0-7864-4292-8
  softcover : 50# alkaline paper

  1. Natchez (Steamboat)   2. Robert E. Lee (Steamboat)
  3. Steamboats — Mississippi River — History — 19th century.
  4. River steamers — Mississippi River — History — 19th century.
  5. Paddle steamers — Mississippi River — History — 19th century.
  6. Marine engineering — Mississippi River Region — History — 19th century.   7. Shipbuilding — Mississippi River Region — History — 19th century.   I. Title.
  VM625.M5P37   2009
  797.12'5 — dc22   2009011919

British Library cataloguing data are available

©2009 Benton Rain Patterson. All rights reserved

No part of this book may be reproduced or transmitted in any form or by any means, electronic or mechanical, including photocopying or recording, or by any information storage and retrieval system, without permission in writing from the publisher.

On the cover: *The Great Mississippi Steamboat Race: From New Orleans to St. Louis, July 1870* (Library of Congress)

Manufactured in the United States of America

McFarland & Company, Inc., Publishers
  Box 611, Jefferson, North Carolina 28640
    www.mcfarlandpub.com

To the memory of
Robert Townsend Patterson,
chief engineer on the
New Orleans steamer *New Camelia*

Table of Contents

Introduction   1

Part One. The Big Event

1 • The Start   3
2 • The Course   19
3 • The Early Going   35

Part Two. The Origins

4 • The Pioneers   49
5 • A Different Kind of Boat   65
6 • Captain Shreve's Design   77
7 • The Proliferation   89

Part Three. The Circumstances

8 • The Sweet Life on the Mississippi   101
9 • The Hard-Working Life   117
10 • Owners and Officers   129
11 • The Perils   145

Part Four. The Outcome

12 • On to Cairo   167
13 • The Fog   178
14 • Celebration in St. Louis   183

Epilogue	194
Chapter Notes	199
Bibliography	203
Index	206

Introduction

"Nothing," the nineteenth-century steamboat historian E.W. Gould asserted, "so much interests the average American as rapid motion, and it is not confined to our nationality altogether either. The fastest sailing vessel, even a merchantman, always got the preference in the early days, if known to excel in speed. Then followed the clipper ships, which excited the admiration of the civilized world because of their speed.

"Steam had no sooner been applied to navigation than the genius of the best mechanical skill was challenged to produce the best results in speed from a combination of steam power and model of vessels.... The principal question to be determined by all who had embarked in steam navigation was how much speed could be obtained."[1]

And what better way was there to demonstrate how much speed could be obtained, to show which was the fastest vessel, than to race the very fastest against each other?

In the days following America's Civil War two of the very fastest steamboats were the *Robert E. Lee* and the *Natchez*, both operating on the lower Mississippi River, each with a large following of customers and friends. The personal rivalry of their owner-captains and the public partisanship that the boats engendered grew so intense that a match between the two became inevitable. The resulting race won both boats a fame so widespread and enduring that no other steamboats would ever equal it. The race itself became so famous that it became a milestone in the history of America.

The *Natchez* and the *Robert E. Lee* were quintessential Mississippi River steamboats, elegant specimens of the breed, built to tempt aboard passengers who could afford to travel in style, much like twenty-first-century cruise ships. Travel aboard a Mississippi River steamboat was, for those who could afford to go first class, an esthetic experience, providing days and nights of pleasure in an opulent floating palace.

But unlike modern cruise ships, Mississippi River steamers had an indispensable practical function, far more important than pleasure or recreation. During many years of the nineteenth century, until the spread of railroads, the steamboat was the major means of transportation for both passengers and freight. The steamboat opened America's mid-continent to settlement by providing access to the roadless western territories, carrying on its often crowded, boisterous main deck those courageous, hardy, sometimes desperate people who settled mid–America, the polyglotinous, multi-ethnic immigrants from abroad as well as restless and hopeful Americans moving westward from states along the eastern seaboard, all seeking new opportunity in a land of opportunity. For them the promise of America lay within its immense interior, which was reachable only on foot, through and across largely trackless woods and plains, or by boats steaming through the growing nation's intricate network of rivers.

The steamboat was the way in and the way out. Once on their land, the settlers, farmers and planters depended on the steamboat to take the fruits of their labors to market centers where they could be sold, and to bring from those market centers what people needed to survive or simply to make their lives better. People of the mid-continent turned the Mississippi River into a vibrant thoroughfare, and the steamboat was the vehicle that traveled upon it, transporting them and their goods. A common sight in communities along the river, the steamboat became an integral part of ordinary life in the nineteenth century.

The *Natchez* and the *Robert E. Lee* were only two of the many, but because of the race they ran and the fame it gained them they have become symbols of all Mississippi River steamboats and of the steamboat's time in history. Their story, their vying for pre-eminence, is not merely the story of two of the thousands of steamboats that plied the Mississippi's muddy waters. It is the story of the Mississippi River steamboat itself, the vital, majestic creature of an American era.

Here is that story, from the beginning.

Part One. The Big Event

• 1 •

The Start

It was the most massive crowd on Canal Street since Mardi Gras, despite the summer heat, which afternoon clouds and a light shower failed to abate. The *Daily Picayune* reporter covering the event observed that the city seemed to empty itself onto the levee, thousands of onlookers thronging to where New Orleans's famous thoroughfare meets the mighty river. The *St. Louis Republican* reporter in town for the event said the levee in the area of Canal Street was so densely packed with people that there was practically a solid human mass from the river back a hundred yards or so to the first row of buildings.

In upper-floor windows, on rooftops and on lacy iron balconies people assembled to watch the spectacle. Seven blocks away from the river, on St. Charles Street, as many as a dozen desperate onlookers climbed atop the dome of the St. Charles Hotel to get a clear view of the expected action. As far as the eye could see and farther, from Canal Street all the way uptown to Carrollton and beyond, eager spectators spread themselves out along the river's edge, standing or sitting, squatting or lying wherever they managed to find viewing space, passing the time with food and drink bought from street vendors, suffering the crush gladly, knowing they were about to witness one of history's great moments.

Some spectators, bent on a close-up view of the racers, boarded steamers that had scheduled special excursions to carry paying customers as far as twenty miles up the river, following the boats as the race proceeded. The steamer *Henry Tate* had moved to an upriver vantage point, carrying on board a load of passengers, who had shelled out a dollar apiece for tickets, and a brass band to further enliven the festive atmosphere. A half dozen or so other steamers had joined the *Henry Tate*, all providing the river's equivalent of ringside seats.

Through the courtesy of the Mississippi Valley Transportation Company, the *Picayune* reporter became one of the passenger-spectators aboard

The New Orleans riverfront in the mid–1800s. By 1860 New Orleans had become the largest export shipping point in the world. In 1870, at the time of the historic steamboat race, it was still the No. 1 steamboat port in the nation (Library of Congress).

the steam tug *Mary Alice*, which, like the other vessels, stood in the river waiting for the race to begin. Looking out over the broad expanse of water and across it toward the clusters of buildings in Algiers on the west bank, and noticing where the river lapped the muddy edge of the east bank nearby, the reporter could see that the river was low, as it had been for the past several days. He reported it at six feet and four inches below the high-water mark set eight years earlier.

Steamboat activity at the New Orleans riverfront on this day was not so bustling as it once had been, during the bygone golden days of Mississippi River steamboats, but activity was not exactly languid. Steamboats — and cotton — had helped New Orleans become, by 1860, the largest export shipping point in the world, and in 1870 it was still the No. 1 steamboat port in the nation. Eight packets — as mail- and passenger-carrying steamboats were called — had arrived in the past twenty-four hours and were docked bow-first into the wharves, side by side, like gigantic animals feeding at a trough. The *Mayflower* and the *Wade Hampton* had come from the Ouachita River, the *Bradish Johnson* from Shreveport, the *Hart Able* and *W.S. Pike* from Bayou Sara, the *John Kilgour* from Vicksburg, and the *Enterprise* and *B.L. Hodge* from the Red River. Other packets, including the *Mary Houston* and the *Great Republic*, having arrived earlier, still lay at the wharf, taking on passengers and freight and due to depart on Saturday.

Three departures were scheduled for this day: the *Robert E. Lee,* which had advertised that it was bound for Louisville, but which no one believed it was; the *Natchez*, bound for St. Louis, as everyone knew; and the *Grand Era,* bound for Greenville, Mississippi. The *Natchez*'s usual run was between New Orleans and St. Louis. The usual run of the *Robert E. Lee* was between New Orleans and Louisville. Ordinarily those two boats never left New Orleans on the same day. But on this day, Thursday, June 30, 1870, they were going to do something out of the ordinary.

The customary departure time for steamboats leaving New Orleans was between four and five P.M., and their leaving invariably created a riverfront scene that, having once been witnessed, remained a vivid impression on those who had experienced it. The onetime steamboat pilot Samuel L. Clemens of Hannibal, Missouri, who quit steamboating and became author Mark Twain, long remembered the sights and sounds of the New Orleans waterfront departure scene and described them for the readers of his classic work, *Life on the Mississippi*:

> From three [P.M.] onward they [the steamboats] would be burning rosin and pitch-pine (the sign of preparation), and so one had the picturesque spectacle of a rank, some two or three miles long, of tall, ascending columns of coal-black smoke; a colonnade which supported a sable roof of the same smoke blended together and spreading abroad over the city.
>
> Every outward-bound boat had its flag flying at the jack-staff, and sometimes a duplicate on the verge-staff astern. Two or three miles of mates were commanding and swearing with more than the usual emphasis; countless processions of freight barrels and boxes were spinning athwart the levee and flying aboard the stage-planks; belated passengers were dodging and skipping among these frantic things, hoping to reach the forecastle companionway alive...; women with reticules and bandboxes were trying to keep up with husbands freighted with carpet sacks and crying babies...; drays and baggage-vans were clattering hither and thither in a wild hurry, every now and then getting blocked and jammed together...; every windlass connected with every fore-hatch, from one end of that long array of steamboats to the other, was keeping up a deafening whizz and whir, lowering freight into the hold, and half-naked crews of perspiring negroes that worked them were roaring such songs as "De Las' Sack! De Las' Sack!"... By this time the hurricane and boiler decks of the steamers would be packed black with passengers. The "last bells" would begin to clang, all down the line...; in a moment or two the final warning came — a simultaneous din of Chinese gongs, with the cry, "All dat ain't goin', please to git asho'!"... People came swarming ashore, overturning excited stragglers that were trying to swarm aboard. One more moment later a long array of stage-planks was being hauled in....
>
> Now a number of the boats slide backward into the stream, leaving wide gaps in the serried rank of steamers.... Steamer after steamer straightens herself up, gathers all her strength, and presently comes swinging by, under a tremendous head of steam, with flag flying, black smoke rolling, and her entire crew of firemen and deck-hands (usually swarthy negroes) massed together on the forecastle ... all roaring a mighty chorus, while the parting cannons boom and the multitudinous spectators wave their hats and huzza! Steamer after steamer falls into line, and the stately procession goes winging its flight up the river.[1]

As five o'clock approached, the clamor of departure on this day seemed even more boisterous than what Clemens remembered. Two of the steamboats about to shove off from the wharf were going to commence the most

One. The Big Event

Steamboats lined up at the New Orleans wharves around 1870. Mark Twain captured the excitement of such New Orleans riverfront scenes in his classic work, *Life on the Mississippi* (Library of Congress).

promoted, most talked about, most speculated over, most gambled on steamboat race in history.

Everyone along the river, in towns, villages and cities and the spaces in between them, had heard about it, as had a great many in cities far from the banks of the Mississippi, across the country and across the seas. The race had captured the attention and imagination of almost everybody. And most of those nestled in the huge crowd of spectators, white and black, employer and employee, rich and poor, man and woman, boy and girl, had a favorite they were pulling for. All were expecting to see the beginning of the race of the century, pitting two of the biggest, speediest and best-known packets against each other, the *Natchez* versus the *Robert E. Lee*, running from New Orleans to St. Louis, twelve hundred river miles, as fast as their huge paddle wheels — and their captains — could drive them.

The early-twentieth-century steamboat historians Herbert and Edward Quick, who lived at a time that was close to America's steamboating era, evinced the feelings of many people of those days:

To those who merely looked on, a steamboat race was a spectacle without an equal. To the people of the lonely plantations on the reaches of the great river, the sight of a race was a fleeting glimpse of the intense life they might never live. To see a well-matched pair of crack steamboats tearing past, foam flying, flames spurting from the tops of blistered stacks, crews and passengers yelling — the man or woman or child of the backwoods who had seen this had a story to tell to grandchildren.[2]

The people of New Orleans, of course, where the race would start, were especially fascinated, even obsessed. The *Picayune* declared, "The whole town is given up to the excitement occasioned by the great race.... Enormous sums of money have been staked here on the result, not only in sporting circles but among those who rarely make a wager. Even the ladies have caught the infection, and gloves and bon bons, without limit, have been bet between them."[3] Among the people of New Orleans the *Natchez* was believed to be the favorite, it being considered a New Orleans boat and its owner being a year-round New Orleans resident.

In other cities along the Mississippi and Ohio interest in the race was almost equally high as in New Orleans. The *New York Times* reported from Memphis that "the excitement over the race between the *R.E. Lee* and the *Natchez* is intense. The betting is heavy, with the odds in favor of the *Lee*." In St. Louis, it reported, "The excitement over the steam-boat race is very great here this evening, and large amounts of money are staked...." And in Cincinnati: "The race between the steamers *Natchez* and *R.E. Lee*, on the Mississippi River, has created more of a sensation here today than anything of the kind that ever occurred. There has been a great deal of betting. Between $100,000 and $200,000 have doubtless been staked."[4]

There is no way of knowing how enormous was the total sum bet on the race, but it easily rose into the millions. Professional gamblers were having a field day. In New Orleans before the race began, they were giving odds on the *Robert E. Lee*. Seventy-five-dollars bet on the *Natchez* would return one hundred dollars if it beat the *Lee*.

More than bets were at stake, though. Winning a head-to-head race, and thereby establishing itself as the fastest steamboat on the river, would be a public relations and marketing windfall, potentially bringing new freight and passenger business to the winner, and increased profits along with it. Losing the race, particularly if by a considerable time, would be a humbling if not humiliating experience for both the boat and its crew, and possibly a costly one in lost future revenue.

Some of the backers of the race, influentials who had helped persuade the *Robert E. Lee*'s reluctant owner and captain, John W. Cannon, to agree to the contest, had still more in mind. The steamboat business was in a state of decline and had been since the Civil War. The cause was railroads, which over their ever-expanding network of lines could carry passengers and freight faster, cheaper and to and from more destinations than could steamboats. Some steamboat owners who could remember the golden years of the 1840s and 50s thought there was a way to bring back the good times, if only shippers and passengers could have their attention diverted from trains back to the elegant floating palaces that steamboats had been before the war. A race between two of the western rivers' best and fastest steamers — a gigantic publicity stunt — just might help the steamboat business gain new friends and regain old ones that had been lost to the railroads.

The most critical element of the race, however, was the heated rivalry between the boats' owner-captains — tall, powerfully built, craggy-faced, fifty-four-year-old Thomas P. Leathers of the *Natchez* and intense, husky, soft-spoken, fifty-year-old John W. Cannon of the *Robert E. Lee*. Both men were Kentucky natives, Leathers having been born in Kenton County, near Covington, Kentucky, and Cannon near Hawesville in Hancock County, on the Ohio River. Both had long experience with steamboats.

At age twenty Leathers had signed on as mate on a Yazoo River steamer, the *Sunflower*, captained by his brother John. In 1840, when he was twenty-four, he and his brother built a steamboat of their own, the *Princess*, which they operated on the Yazoo and later on the Mississippi, running between New Orleans, Natchez and Vicksburg. The brothers soon built two other steamers, *Princess No. 2* and *Princess No. 3*, and prospered with them on

Thomas P. Leathers, owner and captain of the *Natchez*. Gruff, quick-tempered and physically imposing, Leathers had an intimidating presence and was determined to drive the *Robert E. Lee* off the Mississippi River (National Mississippi River Museum and Aquarium, Captain William D. Bowell Sr. River Library).

the Mississippi. In 1845 Leathers built the first of a series of steamers that he named *Natchez*, each larger and faster than the previous one.

The third *Natchez*, large enough to carry four thousand bales of cotton, met with tragedy when a wharf fire engulfed it and destroyed it, taking the life of Leathers's brother James, who was asleep in his stateroom. The fifth *Natchez*, capable of carrying five thousand bales of cotton, was the boat that transported Jefferson Davis to Montgomery, Alabama, where he was sworn in as the Confederacy's president in 1861. It was operated by Leathers until it was pressed into service by the Confederates, first as a troop carrier and then as a cotton-clad gunboat on the Yazoo River, its works shielded by a wall of cotton bales. On March 23, 1863, twenty-five miles above Yazoo City, Mississippi, it was set ablaze and destroyed by its crew to prevent its capture by Union forces.

Following the fall of New Orleans in 1862, Leathers temporarily gave up the steamboat business. After the war, he returned to the river and in 1869 launched from its Cincinnati shipyard a brand-new *Natchez*, the sixth, of which he was immensely, even overbearingly, proud. He was confident — and boastful — that it could beat anything on the Mississippi.

When in his twenties, he had made his home in Natchez and there he had met Julia Bell, the daughter of a steamboatman, and he married her in 1844, when he was twenty-eight. Julia became a victim of yellow fever, and Leathers had then married Charlotte Celeste Claiborne of New Orleans, member of a prominent Louisiana family that included a former governor, William C.C. Claiborne. Leathers moved from Natchez and made his home in New Orleans, where he and Charlotte began raising a family and where he spent the Civil War years.

Gruff, hard-faced, quick-tempered and physically imposing, Leathers could be intimidating to his workers and to others. Once a steamboat mate, he never got over the use of the profanity-filled language that mates routinely used to drive their crews. Through his years as owner and captain, his mate's vocabulary never left him, and along the river he became notorious for it.

Somewhere during his career Leathers picked up the nickname of "Ol' Push," which according to one account was a shortened form of the name of the heroic, nineteenth-century Natchez Indian chief Pushmataha, whom Leathers and his crew sort of adopted as the symbol and mascot of the *Natchez*, which they liked to call "the big Injun." The nickname, however, could just as easily have been inspired by Leathers's pushy personality. With his fast new *Natchez* he developed the irksome habit of letting other steamers shove off from the New Orleans wharves ahead of him, then while his excited and cheering passengers watched along the rails, he would make a grand show of speeding up and overtaking whatever lesser vessel had moved out into the river before him.

Leathers pulled that stunt once on John Cannon's good friend John Tobin, when Tobin was master of the steamer *Ed Richardson*. Tobin never forgot the incident. He got his chance for revenge when he became captain of the third *J.M. White*, a big, new boat that had never been tested in a race. On a day that Tobin had been waiting for, the *Natchez* and the *J.M. White* were together at the New Orleans waterfront and backed off from the wharf about the same time. The speedy *Natchez* quickly moved out ahead and gained a lead while an accident aboard the *J.M. White* forced it to slow down so that repairs could be made. Once the repairs were completed, Tobin ordered the steam up, and the *J.M. White*, its powerful wheels churning against the current of the muddy Mississippi, glided abreast of the *Natchez*, then overtook it. Leathers, seeing he was beaten, pretended he needed to make a stop to unload freight and thus had to drop out of the contest. The freight that he unloaded was an empty barrel, which he reportedly kept aboard the *Natchez* to be used for just such embarrassing occasions.

Leathers's attitude toward his steamboat business, which he managed with meticulous care, and his position in life were revealed in a story told about him by one of his fellow captains, Billy Jones of Vicksburg. Leathers, Jones claimed, would often refuse to accept the freight for a shipper or a consignee he didn't like, and the firm of Lamkin and Eggleston, a wholesale grocery company in Vicksburg, was one of the shippers he didn't like. When he declined to accept their freight, the firm sued him in circuit court and won a judgment against him. The judgment was upheld in the state supreme court, and Leathers had to pay the firm $2,500 in damages, which infuriated him. "What's the use of being a steamboat captain," he fumed in frustration, "if you can't tell people to go to hell?"[5]

John Cannon, in personality, attitude and some other ways, was completely unlike Tom Leathers. Placid-faced, calm, quiet, he was a careful and far-sighted businessman who seemed more interested in the safety of his boat and passengers than in a showy display or establishing grounds for boasting. But like Leathers, he was a Kentucky farm boy who determined he would make something of himself. As a youngster he paid for his education with money earned by splitting rails. He began his life on the Mississippi aboard a flatboat and deciding the river was where he would pursue his fortune, he became a deckhand on a Red River boat and later a cub pilot on a Ouachita River steamer, the *Diana*, paying his pilot tutor out of the wages he earned working at a variety of jobs aboard the boat.

In 1840 he completed his training and became a licensed pilot. With money he saved from his pilot's pay and with the help of several friends he built the steamer *Louisiana*, which came to a tragic end on November 15, 1849, when its boilers exploded at the Gravier Street wharf in New Orleans, taking eighty-six lives and shattering the two steamers docked on either side of the *Louisiana*. That experience likely affected his way of thinking about endangering other boats of his. Recovering from that disaster, Cannon went on to build or buy a dozen or more steamers, including the *S.W. Downs, Bella Donna, W.W. Farmer, General Quitman, Vicksburg, J.W. Cannon, Ed Richardson* and the *Robert E. Lee*.

The *Robert E. Lee* was built for Cannon in New Albany, Indiana, in 1866, the year after the end of the Civil War, and it was designed to be the most luxurious and fastest boat on the western rivers. When it came time to paint its name on the boat, an explosive problem arose. The display of the name of the South's most famous general inflamed many of the people in New Albany and elsewhere, and Cannon had to have the unfinished boat towed across the Ohio River to the Kentucky side to prevent its being burned by enraged Indiana citizens, whose feelings were expressed in an editorial published in the Rising Sun, Indiana, newspaper, the *Record*, shortly before the race: "The people hereabouts who are interested in the race are friendly to the Natchez for many reasons. A steamer named for any accursed rebel General should scarcely be allowed to float, much less have the honor of making the best time...."

John W. Cannon, owner and captain of the *Robert E. Lee*. Cannon was goaded into racing the *Lee* against the *Natchez* by Tom Leathers, his business rival and personal adversary, with whom he had tangled in a fistfight in a New Orleans saloon.

The *Robert E. Lee* docked beside the *Great Republic*. The *Lee* was built for John Cannon in New Albany, Indiana, in 1866, the year after the Civil War ended. When its name was painted on its wheel housing, the boat had to be towed to the Kentucky side of the Ohio River to prevent its being burned by irate Northerners who objected to its being named for the Confederacy's most famous general (Library of Congress).

Actually, Cannon was believed to have been sympathetic to the Union, though after nearly a lifetime in the steamboat business, he, like Leathers, had many friends and business associates in both the North and South. Cannon was reported to be a friend of Union general Grant, and some suspected that he named his boat after the Confederate general to win approval in the South, where most of his customers were, and to compensate for his known Northern sympathies. Leathers, who refused to fly a U.S. flag on the *Natchez* even though the war had ended and who effected a sort of uniform of Confederate gray while captaining his boat, had at one time been arrested for suspected Union sympathies during the war, only to be pardoned by his friend Jefferson Davis, the Confederate president and former United States senator from Mississippi.

Cannon had managed to make a small fortune out of the war. He took his steamer *General Quitman* up the Red River and kept it hidden until Union forces had taken complete control over the Mississippi, then came steaming down from Shreveport to the Mississippi and up to St. Louis with a boatload of cotton that he had bought at depressed prices from planters unable to sell it on the usual markets. He then sold it for several times the price he had paid, netting a profit estimated at $250,000. With that bankroll, Cannon had little trouble paying the $230,000 that the *Robert E. Lee* cost him,[6] or being able to afford two homes, one in New Orleans and the other in Frankfort, Kentucky, where he and his wife spent their summers.

The completed *Robert E. Lee* arrived in New Orleans on its maiden voyage in October 1866 and quickly proved itself as a fast steamer, setting new speed records and winning over flocks of new customers — while at the same time raising the ire and jealousy of Tom Leathers, who for a short time after the war had worked for Cannon as captain of the *General Quitman*. Cannon was said to have taken pleasure in having his rival work for him. The relationship between the two men remained a stormy one, though, and Leathers had left the *General Quitman* to command another steamer, the *Belle Lee*, in 1868.

In November 1868 their hostility toward each other broke out in a quarrel in a New Orleans saloon, while both apparently were under the influence, and a fistfight resulted. "Very little claret was drawn," a waggish reporter commented while declaring that "Capt. Cannon had the best of the fight."[7] After that scuffle, they refused to speak to each other or even to exchange whistle salutes as their boats passed each other on the river, which was the custom among steamboatmen.

Reserved though he was, Cannon was not one to back away from a fight. In 1858, when he was master of the *Vicksburg*, he fired a pilot named Allen Pell and when Pell demanded to know the reason, Cannon, apparently in no uncertain terms, told him. The blunt answer drew an angry threat from Pell, and Cannon, thick-bodied and strong-armed, drove a heavy fist into Pell's face in reaction, staggering him. Pell then pulled a knife from the sleeve of his coat, and Cannon, undaunted, grappled with Pell for the knife and was stabbed just above the groin. Cannon recovered and although his dealings with Pell were forever ended, Cannon had no qualms about hiring one of Pell's relatives, James Pell, as one of his pilots aboard the *Robert E. Lee*.

In 1869, determined to best Cannon and boasting that he would drive the *Robert E. Lee* off the river, Leathers ordered a shipyard in Cincinnati to build him a boat, a new *Natchez*, that would outrace Cannon's speedy, elegant *Robert E. Lee*. Leathers had come out of the war not nearly so well off

The *Natchez*, the sixth steamer to bear that name. It was built for Tom Leathers in Cincinnati in 1869 at a cost of $200,000 and was designed, by Leathers' orders, to outrun the swift *Robert E. Lee* (National Mississippi River Museum and Aquarium, Captain William D. Bowell Sr. River Library).

as Cannon, having lost his boat during the conflict. He found financial backers, however, the chief of whom was Cincinnati businessman Charles Kilgour. The new *Natchez*, completed at a cost of some $200,000, began its first voyage down the Ohio and into the Mississippi on October 3, 1869. In June 1870 he was said to still owe $90,000 on the boat.

Cannon had repeatedly declined challenges to pit his boat against the *Natchez*. For a while, ever since the *Lee* had first been put into service, Cannon had run it between New Orleans and Vicksburg, the same run that the *Natchez* made. The *Lee* would leave New Orleans every week on Tuesday and the *Natchez* on Saturday. Fans of the two boats developed a sense of rivalry between them, each group sure of the superiority of their favorite and urging the two captains to race them and settle the question of which was faster. As if to dampen the enthusiasm for a race, Cannon altered the *Robert E. Lee*'s run in the spring of 1870. Its new schedule had it run between New Orleans and Louisville, leaving New Orleans every other Thursday. Leathers also dropped out of the New Orleans–Vicksburg trade that spring and he began running the *Natchez* between New Orleans and St. Louis, leaving New Orleans on Saturdays.

Calls for a race increased despite the change in schedules. Leathers was all for it. Throwing down his gauntlet, Ol' Push announced that the *Natchez*, instead of leaving New Orleans on Saturday as usual, would leave on the same day, at the same time, that the *Robert E. Lee* left, forcing Cannon to run the *Lee* against him. Cannon still resisted, but on his return trip from Louisville in late June his shippers along the Ohio River repeatedly urged him to race. At last he gave in to the pressures coming from customers, from newspapers in the river towns from New Orleans to Louisville, from planters and merchants and other businessmen, from gambling interests — and from Leathers himself. Still, after he had agreed to the contest, he denied reports that he would engage in the race and had a notice to that effect published in successive editions of the *Picayune*, including the edition published on the morning of the day the race was to begin:

A CARD

Reports having been circulated that steamer R.E. LEE, leaving for Louisville on the 30th June, is going out for a race, such reports are not true, and the travelling community are assured that every attention will be given to the safety and comfort of passengers.

The running and management of the Lee will in no manner be affected by the departure of other boats.

John W. Cannon, Master

Leathers, likewise apparently fearing repercussions from some passengers and shippers, also had a denial published in the *Picayune*:

A CARD TO THE PUBLIC

Being satisfied that the steamer NATCHEZ has a reputation of being fast, I take this method of informing the public that the reports of the Natchez leaving here next Thursday, the 30th inst., intending racing, are not true.

All passengers and shippers can rest assured that the Natchez will not race with any boat that will leave here on the same day with her. All business intrusted to my care, either in freight or passengers, will have the best attention.

T.P. Leathers,
Master, Steamer Natchez

The *Robert E. Lee* arrived back in New Orleans from Louisville in the evening of Tuesday, June 28, and Cannon, having by then agreed to the race, having begun elaborate preparations for it and having become determined that his hated rival would not beat him, immediately ordered his magnificent *Lee* stripped of every possible impediment to speed. To reduce wind resistance, the window sash was removed from the front and back of the pilothouse, and the front double doors and the big aft windows of the main cabin were likewise removed. The decks aft of the paddle wheels had every other plank taken up to allow the spray from the wheels to quickly fall through the decks. The steam-escape pipes, the freight-lifting derricks, the spare anchors and extra mooring chains, everything that could be spared on the main deck and in the hold was taken ashore, along with virtually everything else that was portable, including most of the staterooms' furniture and decorative accessories, and all freight had been refused. Left in place, however, was the large, handsome portrait of the boat's namesake, General Lee.

The passenger list had been kept as short as possible. Captain Cannon announced to passengers who already held tickets that plans had changed and the *Lee* was headed for St. Louis, not Louisville, and there would be no stops on the way. Passengers who had bought passage to Louisville and other destinations on the Ohio would be transferred to another steamer at Cairo, Illinois.

Even so, Cannon would have to accommodate some seventy passengers, including friends and some fellow captains, business associates and VIPs, all of them presumably eager to be participants in the history-making race. Two who perhaps were not so eager were the twenty-six-year-old carpetbagger governor of Louisiana, Henry Clay Warmoth, and his close friend and political ally, Doctor A.W. Smyth, chief surgeon at Charity Hospital in New Orleans. Smyth was also a close friend of John Cannon, and the two men — Smyth and Warmoth — had just arrived at the New Orleans riverfront on a steamer from Baton Rouge, where Warmoth had presented diplomas to graduates at Louisiana State University's commencement exercises. What the two men encountered when they reached the wharf was the largest crowd the governor had ever seen in New Orleans. On the spur of the moment, Smyth had suggested they board the *Robert E. Lee* to greet Captain Cannon and wish him well. Delighted to have them aboard, Cannon insisted they stay for the ride. "Captain Cannon pressed us to go with him," Warmoth wrote later, the event evidently a memorable one, "and, as we were carried away by the excitement and enthusiasm, we accepted the invitation."[8]

Leathers, supremely confident of the *Natchez*'s prowess, had made no such preparations. The only concession he had made was the removal of the boat's landing stage that swung from a derrick and which he acknowledged could catch the wind and slow the vessel somewhat. He had booked ninety passengers aboard the *Natchez*, with destinations requiring stops at Natchez, Vicksburg, Greenville, Memphis and Cairo. Others intending to board the boat would be waiting for it along the levee upriver from New Orleans. Leathers had also taken on a load of freight, evidently considering this run to St. Louis to be business as usual, only made at a greater speed than his rival who, he apparently believed, would also make a more or less normal trip.

The *Natchez*'s freight and passenger load would add considerable weight to the vessel, but despite it, Leathers's boat would draw but six and a half feet of water, a foot less than the stripped and lightened *Robert E. Lee*. The difference in draft could be important in the race, and not only because the shallower-draft vessel would meet less resistance in the water. The Mississippi was reported to be falling, increasing the danger of a deep-draft boat's running aground on the river bottom.

Other than their draft and a difference in their length and freight capacity, the two steamers, both side-wheelers, were about equal in size and equipment. The *Robert E. Lee* was 285 feet in length and forty-six feet in the beam; the *Natchez* was 303 feet long and forty-six feet in the beam. The height of the *Robert E. Lee* to its pilothouse was thirty and a half feet; the height of the *Natchez* was thirty-three feet. The *Robert E. Lee*'s paddle wheels were thirty-eight feet in diameter and seventeen feet wide. The *Natchez*'s paddle wheels were forty-two feet in diameter and eleven feet wide. Each boat had eight boilers, the *Natchez*'s being slightly larger (thirty-three feet long) than the *Robert E. Lee*'s (twenty-eight feet) and capable of generating higher pressure (160 pounds) than the *Robert E. Lee*'s boilers (110 pounds). The engines were also similar. The *Robert E. Lee*'s cylinders were forty inches in diameter with a ten-foot stroke; the *Natchez*'s cylinders were thirty-six inches in diameter with a ten-foot stroke.[9]

To make sure, as sure as could be made, that the *Lee* was complying—and would continue to do so—with steamboat safety regulations, a U.S. steamboat inspector, a man named Whitmore, came aboard the vessel and examined the safety valves on each of the eight boilers. On each valve that could be locked, after locking it, he placed the government's lead seal. Engineers would not then be tempted to manipulate the safety valves to increase steam pressure.

At fifteen minutes to five P.M., a quarter hour before the announced departure time for both boats, Captain Cannon gave three tugs on the *Robert E. Lee*'s ship's bell to signal it was time for visitors to hurry ashore and for passengers to find their staterooms or a place at the rails. Captain Leathers immediately followed with three clangs on the *Natchez*'s bell. The last of the visitors having hustled ashore, the *Lee*'s mate shouted the order for the landing stage to be hauled in. As thick, black smoke erupted from the *Robert E. Lee*'s soaring chimneys, its bell sounded once more, an axe blade fell and severed the bowline that bound the vessel to the wharf, and the axe wielder suddenly raced for the end of the landing stage, grabbed it and held on as it came sliding onto the main deck.

Instantly then, minutes short of five o'clock, the big, grand vessel, its white woodwork gleaming in the afternoon sun, moved stern first into the streaming current of the Mississippi River, its giant paddle wheels churning a froth in the muddy flow.

The slightly early start had been carefully arranged by Cannon. He had gathered his officers together at four o'clock and given them instructions which, according to one of the assistant engineers, John Wiest, went as follows: "I want everybody aboard at five o'clock. The pilot in his house, but not in sight, the engineers at the throttle valves, the mate to have only one stage out and that at a balance so that the weight of one man on the boat end will lift it clear of the wharf. There will be a single line out, fast to a ring bolt, with a man stationed there, axe in hand, to cut and run for the end of the stage the moment he hears a single tap of the bell, and come aboard on the run or get left."[10]

Knowing Leathers's reputation for making sudden fast starts against competitors, Cannon had now out-fast-started him. The *Lee* had been docked just below the *Natchez*, and as it backed out from the wharf and made a crescent-shaped turn to head its bow upriver, the *Natchez* was forced to wait for Cannon to straighten out the *Lee*, lest the *Natchez* back across the *Lee*'s bow, or possibly into it. Once headed upstream, the *Robert E. Lee* fired its signal cannon as it passed St. Mary's Market, just above Canal Street, the official starting point for timing all steamboat voyages from New Orleans. The time was a minute and forty-five seconds before five o'clock.[11]

As soon as the *Lee* had moved out of its way, the *Natchez*, distinguishable from a distance by its bright-red chimneys, backed away from the wharf, straightened out and with a surge of power steamed up to St. Mary's Market, firing its signal cannon as it passed, the gun's deep *boom* resounding over the noise of the yelling crowds on both sides of the river. The time then was two minutes after five o'clock.[12]

The race was on. Twelve hundred miles of river lay ahead.

· 2 ·

The Course

Alonso Álvarez de Piñeda was the first known European to see the big river. His view of it came from across the rail of a sailing ship as he and his Spanish exploration fleet followed the coast of the Gulf of Mexico in 1519, sailing from Florida, bound for Mexico. Around June 2 he passed the outflow of a mighty stream and he made a note of it in his log and gave it a name — Rio del Espiritu Santo, or Holy Spirit River, because he had sighted it on (or around) the Catholic feast day of Espirito Santo, or Pentecost. But noting and naming it constituted the extent of Álvarez's interest in the river, and he sailed on to Cabo Rojo, Mexico.

Another Spaniard, Álvar Núñez Cabeza de Vaca, was the next European to see it. His view was a lot closer up than Álvarez's. He and his contingent of explorers were traveling cross-country, slogging and plodding their way from Florida toward Mexico, exhausting themselves in the wilds and marshes, amid harassing Indians in Louisiana and Texas. Sometime in 1528 Cabeza de Vaca and his party managed to cross the Mississippi near its mouth and kept going on what was to become an eight-year trek, which only Cabeza de Vaca and three others of his party survived.

Hernando de Soto, another Spanish explorer, made deadly enemies of the Indians in Florida, Alabama and Mississippi and was forced to fight them repeatedly as he and his steadily diminishing army, starting at Tampa Bay, moved up the Florida peninsula into Georgia and South Carolina, then turned west and traversed Alabama and Mississippi, coming at last, perhaps near present-day Greenville, Mississippi, to the banks of the wide and muddy river, which he saw merely as an obstacle on his way to the imagined gold that awaited his looting. He crossed the river and traveled as far as the northwest part of Arkansas before turning around and heading back to the big river. After three years of fruitless searching for treasure, he contracted a disease and died near the banks of the river in June 1542. The few survivors of

Père Jacques Marquette and Louis Joliet descend the Mississippi River by canoe in 1673. They were the first Europeans to explore the river, seeking to discover where it would take them. Marquette recorded in his journal the name the natives gave the river — Missisipi, as Marquette spelled it, meaning "great river" (Library of Congress).

his army placed his corpse in a hollowed-out tree trunk and sank it in the river, lest the Indians discover that despite what he had told them, he was not immortal after all. The survivors then fled down the river and made their way to Mexico.

A 35-year-old French Jesuit missionary priest, Jacques Marquette — known to history as *Père* Marquette — and a Canadian-born Jesuit seminary dropout turned explorer, 27-year-old Louis Joliet (or Jolliet), were the first Europeans to actually explore the big river. Unlike the Spaniards who had preceded them, Marquette and Joliet set out not merely to cross the mighty stream but to discover where it would take them — perhaps, they thought, to the Pacific Ocean. Further unlike the Spaniards, they meant to befriend and proselytize the natives they would meet along their way, not conquer and pillage them. It was Marquette, moreover, who recorded in his journal the name the natives had given the big river — Missisipi, as Marquette spelled it, the "great river."

2 • The Course

On May 17, 1673, Marquette and Joliet launched their two canoes into Lake Huron near Michelmackinaw, at the lake's western end, and with five fellow explorers, all of them half Indian and half French Canadian, began paddling their way westward through the Straits of Mackinac and into Lake Michigan, then along the south shore of Michigan's Upper Peninsula. Arriving at the mouth of the Fox River at present-day Green Bay, Wisconsin, Marquette and Joliet ascended the Fox to Lake Winnebago, then continued on to a Mascouten Indian settlement at present-day Berlin, Wisconsin, where they stopped and rested and learned of a portage, farther south, that would lead them to the Wisconsin River, which flows into the Mississippi. Following the Indians' directions, they crossed overland to the Wisconsin River near Portage, Wisconsin, put their canoes back into the water and traveled downstream to the Wisconsin's confluence with the Mississippi near Prairie du Chien, reaching the big river on June 17, 1673, exactly a month after their departure.

Days of further travel led Marquette to deduce that, contrary to their hopes, they were not headed toward the Pacific Ocean. "Judging from The Direction of the course of the Missisipi," he wrote, "if it Continue the same way, we think that it discharges into the mexican gulf." When the explorers passed what apparently was the mouth of the Missouri River, however, Marquette guessed that that was the river that would take them toward California and in his journal he expressed the hope that he might later explore it.

Marquette and Joliet continued their brave descent of the Mississippi as far as the Arkansas River and there they decided they had gone far enough. They estimated that they were within a few days of the Gulf of Mexico and were fast approaching Spanish-held territory and consequently feared capture or worse.

And so on July 17, 1673, the little party of explorers turned their canoes upriver and started the trip back to the French settlements whence they had come, finally arriving in late 1674 after making several stops along the way. Marquette never got to explore the Missouri. He died in 1675, a victim of dysentery contracted on his historic voyage down the Mississippi. Joliet married shortly after his return. In 1680, as a reward for his service to Canada (or New France), Joliet was granted Anticosti Island, at the mouth of the St. Lawrence River. In May 1700 he became lost and died while on an expedition to one of his land holdings.

Rene-Robert Cavelier, *sieur* de La Salle, born in Rouen, France, in 1643, was perhaps the first to see the Mississippi for what it was — a broad highway that, with its tributaries, provided access into the heart of a vast continent teeming with promise and potential. Another dropout from the Jesuit priesthood, he left France to seek a new life in Canada in the spring of 1666, arriving in Quebec in 1667. He managed to acquire a land holding on the western end of the island of Montreal, a section known as Lachine. From the Iroquois natives in the area, whose language he learned, La Salle heard stories of a river called the Ohio that flowed into the great river, the Mississippi. Without the benefit of what *Père* Marquette and Louis Joliet were later to discover, La Salle leaped to the conclusion that the Mississippi was the hoped-for route to the Pacific Ocean and China and started making plans to explore it.

With a go-ahead from the Canadian governor and after selling Lachine to finance his expedition, La Salle in 1669 set out for the Ohio with a party of fifteen men in five canoes. He claimed to have reached the Ohio and to have followed it as far as present-day Louisville, but he didn't make it to the Mississippi. His attention was diverted to the establishment of a fur-trading business, at which he became a success. In 1682, apparently bored with the fur business, he had another go at exploring the Mississippi. He launched an expedition of twenty-three Frenchmen and eighteen Indians from Fort Crevecoeur, near present-day Peoria, Illinois, into canoes and descended the Illinois River to the Mississippi and then paddled down the big river to present-day Memphis, where he built a fortification he named Fort Prudhomme. From Memphis he and his party continued down the river all the way to the Gulf of Mexico, stopping at a site near present-day Venice, Louisiana, on April 9, 1682, to plant a marker post and a cross that claimed for France the entire Mississippi River basin, including all the land drained by the big river and its many tributaries. Nineteenth-century historian Francis Parkman vividly memorialized the momentous event:

> On that day the realm of France received ... a stupendous accession. The fertile plains of Texas; the vast basin of the Mississippi, from its frozen northern springs to the sultry borders of the Gulf; from the woody ridges of the Alleghanies to the bare peaks of the Rocky Mountains — a region of savannas and forests, sun-cracked deserts and grassy prairies, watered by a thousand rivers, ranged by a thousand warlike tribes, passed beneath the scepter of the Sultan of Versailles; and all by virtue of a feeble human voice, inaudible a half a mile.[1]

And to that immense, diverse territory La Salle gave a name. He called it Louisiana, in honor of the king of France, Louis XIV.

From the mouth of the Mississippi La Salle returned to Canada and then to France. In 1684 he again left France, this time with four ships and three hundred hopeful emigrants to found a colony on the Gulf of Mexico. The venture suffered a series of disasters, including attacks by pirates and Indians and the woeful consequences of poor navigation that took them farther west than they apparently intended to go. One of the ships was lost to pirates in the West Indies, another sank in an inlet off Matagorda Bay, and the third ran aground at Matagorda Bay. La Salle and the other survivors erected a fortification near present-day Victoria, Texas, and La Salle then led a group on foot to seek out the Mississippi River, a futile effort that ended when the thirty-six surviving members of the expedition mutinied. Four of the mutineers murdered La Salle on March 20, 1687, near present-day Navasota, Texas. The little colony that he had planted was wiped out in 1688 when Indians slaughtered the colony's twenty adults and carried off their five children as captives.

The intrepid explorers of the Mississippi during the sixteenth and seventeenth centuries demonstrated that the big river flowed uninterrupted from high in the continent's heartland through changing climes to the Gulf of Mexico, the mid-continent's gateway to the seven seas. But the early explorers never traced the Mississippi northward to its source. That notable deed awaited the coming of Henry Rowe Schoolcraft.

Restless and curious, Schoolcraft disdained staying near home, joining his family's glassmaking business and leading a conventional life. In 1818, twenty-five years old and still single, he bade his family and friends in Albany, New York, his hometown, goodbye and set off on a journey of exploration that would let him follow his interests in geography, geology and mineralogy. In 1821 he joined an expedition led by General Lewis Cass, probing the upper peninsula of Michigan and northern Minnesota and hoping to discover, among other things, the source of the Mississippi. In Minnesota Cass and his party of explorers found a lake they decided was the river's headwaters and named it, as something of a memorial, Cass Lake.

Back from that adventure, Schoolcraft took a job as an Indian agent in 1822, stationed at Sault Ste. Marie, Michigan, at the northeastern tip of the upper peninsula. There he met and married Jane Johnson, daughter of an Irish fur trader and an Ojibway woman, and from her he learned a great deal about Indian culture and language. In 1832, on a mission to smooth over relations between the quarreling Chippewas and the Sioux, he determined that the Mississippi did not originate at Cass Lake and decided to try to find the big river's true source.

After days of paddling upstream and across lakes and portaging through sandy, brushy, marshy and piney wilderness, Schoolcraft's party of explorers discovered that the stream of the Mississippi separated into two branches above Cass Lake, something that the available maps failed to show.

The explorers pressed on, wearied by the demands of the portage and stopping often to rest and lay down their burdens for brief respites, and at last came the accomplishment of their arduous mission, recounted by Schoolcraft in his journal:

> Every step we made in treading these sandy elevations, seemed to increase the ardor with which we were carried forward. The desire of reaching the actual

source of a stream so celebrated as the Mississippi — a stream which La Salle had reached the mouth of, a century and a half (lacking a year) before, was perhaps predominant; and we followed our guide down the sides of the last elevation, with the expectation of momentarily reaching the goal of our journey. What had been long sought, at last appeared suddenly. On turning out of a thicket, into a small weedy opening, the cheering sight of a transparent body of water burst upon our view. It was Itasca Lake — the source of the Mississippi.[2]

Known to the French as Lac la Biche, the lake, as described by Schoolcraft, was "a beautiful sheet of water, seven or eight miles in extent, lying among hills of diluvial formation, surmounted with pines, which fringe the distant horizon and form an agreeable contrast with the greener foliage of its immediate shores." The outlet of the lake, through which it begets the Mississippi River, was ten to twelve feet wide, and the water there, as it poured into a stream, was twelve to eighteen inches deep. From such a beginning came the mighty Mississippi.

Schoolcraft gave the lake a new name, one that he contrived by splicing together parts of two Latin words, "*veritas caput,*" which translate into English as "true head"— meaning the river's actual source. Thus the lake became Lake Itasca.

Schoolcraft's discovery of the "true head" provided the Mississippi's total measurement, from source to finish. From Lake Itasca in Minnesota the river stretches approximately 2,350 twisting, curving miles to its debouchment into the Gulf of Mexico, its course and length forever changing with the vagaries of its flow. It receives into its broad stream the waters of some 250 tribu-

Henry Rowe Schoolcraft, discoverer of the source of the Mississippi in 1832. He named the Minnesota lake from which the river sprang Lake Itasca, a name he coined by splicing together parts of the Latin phrase "*veritas caput,*" meaning "true head" (Library of Congress).

taries, and the area that it drains comprises about 1,250,000 square miles, nearly half of the continental United States.

The Mississippi is America's mightiest river — and its most important, a fact keenly realized by President Thomas Jefferson and his secretary of state, James Madison, who purposed to gain the free navigation of the river and acquire for the United States the city that commanded the river's outlet to the sea. "There is on the globe," Jefferson wrote in 1802, "one single spot, the possessor of which is our natural and habitual enemy. It is New Orleans, through which the produce of three-eighths of our territory must pass to market, and from its fertility it will ere long yield more than half our whole produce and more than half our inhabitants."[3]

"The navigation of the Mississippi," President Jefferson declared, "we must have."[4]

Control of the Mississippi and access to the Gulf of Mexico were, together, a highly inflammatory issue in 1802. France had lost much of its valued New World territory as the price of peace in the French and Indian War, which ended in 1763, but it still had aspirations of empire in America. France had undergone a revolution beginning in 1789, which had deposed Louis XVI and swept away most of the old order, and in late 1799 General Napoleon Bonaparte in a fraudulent popular election had been voted first consul of the newborn French republic and had taken over the French government. On March 21, 1801, he had re-acquired from Spain the vast Louisiana territory as a first step in his plan for French expansion. But in the spring of 1803 he changed his mind, his thoughts shifting away from the New World and settling instead on the nearby hated nation that stood as the major obstacle to his achievement of world domination. The conflict he sought and the conquest he desired were not in America, he decided, but rather in England. Louisiana became disposable.

After several tough bargaining sessions, Robert R. Livingston and James Monroe, representing the United States, bought Louisiana, including New Orleans, for about twenty million American dollars. They signed the purchase agreement on May 2, 1803, in Paris.

All concerned were delighted. "The negotiations leave me nothing to wish for," Napoleon remarked. Monroe grandly called the negotiations the "extraordinary movements of the epoch in which we live." Perhaps seeing much farther than the others, Livingston exultantly declared, "This is the noblest work of our whole lives."

By a vote of twenty-four to seven, the United States Senate on Monday, October 17, 1803, ratified the treaty of purchase, the final action needed to seal the deal. By its extraordinary purchase the United States acquired an

View of New Orleans in 1839. The United States acquired the city in the Louisiana Purchase in 1803. Its importance was emphasized by President Thomas Jefferson in 1802. Through New Orleans, he wrote, "the produce of three eighths of our territory must pass to market." New Orleans controlled navigation on the Mississippi, and, "The navigation of the Mississippi," Jefferson declared, "we must have" (Library of Congress).

additional 827,987 square miles, or 529,911,681 acres, more than 22 percent of the present-day United States, everything from the Mississippi River to the Rocky Mountains, including all or parts of the states of Arkansas, Missouri, Iowa, Minnesota, North Dakota, South Dakota, Nebraska, Kansas, Oklahoma, Texas, New Mexico, Montana, Wyoming and Colorado as well as the state of Louisiana.

Not only the Mississippi River but all the land that it drained, from the east and from the west, would now forever belong to the United States and its people. A whole new era of American agriculture, industry, commerce and transportation had dawned, brilliant with opportunity and promise. All that was needed then was a suitable vessel to travel the big river, bearing settlers and developers into the mid-continent's rugged vastness and carrying from it the wealth of produce, materials and products that it yielded.

Travel *down* the big river was not much of a problem. Canoes and pirogues and, later, flatboats, keel boats and barges simply went with the current, steered by sturdy river boatmen manning paddles or oars. Travel *up* the river was another matter entirely. The river's many bends and twists rendered sail power impractical, so that boats going upstream had to depend on the manpower of their crews, who laboriously poled, paddled, rowed or towed their vessels against the relentless current to upriver destinations. Those arduous and limited methods of propulsion prevented the full use of the river and stood in the way of America's realization of its tremendous potential.

Then came a revolutionary, history-changing invention. Many men, both in the United States and in Europe, contributed to its development, but it was Robert Fulton, a poor immigrant's son from Pennsylvania, who made it work successfully. To him went the credit and the fame for the creation of the steamboat.

No longer then was the river master. It became servant. Perspicacious witnesses to the coming of the earliest steamboats realized what was happening. When the first steamer to ply the Mississippi, the *New Orleans*, pulled into Natchez on its maiden voyage in January 1812, an elderly slave who watched it in admiration immediately sensed its meaning. Throwing his hat into the air, he exultantly shouted, "Ol' Mississip done got her master now!" Or so the story goes.

Development of the land and resources along the Mississippi and its tributaries rapidly followed. The banks of the river, on both sides, became dotted with settlements and towns and the landings for the steamboats that were the main means of transportation. New communities sprang up, and older ones grew larger and busier. Travelers on the steamboats that served the river cities and towns got sort of a water bird's view of the mid-continent from the decks of the boats. For many, particularly nineteenth-century immigrants, the voyage into America's heartland began at the city that was founded to serve as the mid-continent's gateway.

It stood as a geographic curiosity, perilously poised on the east bank of the threatening river, the storied city of New Orleans, which by the middle of the nineteenth century had become the commercial terminus of the vast Mississippi valley. From the steamboats' upper decks passengers could peer down on the city, over the ridge of the protective levee, viewing the city's structures as if from an elevated railway, which was the sight that onetime river pilot Samuel Clemens remembered seeing as his vessel approached the city. "In high-river stage, in the New Orleans region," he wrote, "the water is up to the top of the inclosing levee rim, the flat country behind it lies low — representing the bottom of a dish — and as the boat swims along, high on the flood, one looks down upon the houses and into the upper windows. There is nothing but that frail breastwork of earth between the people and destruction."[5]

As steamers bucked the muddy flow and churned northward from New Orleans they made stops where their freight or passengers required, often being hailed to shore by passengers seeking to board them from an isolated spot on the levee. But many of their landings were regular stops, one of the first of which, going upriver, was Donaldsonville, Louisiana, where Bayou Lafourche — which a couple of millennia or so ago was the main stream of the river — splits off from the Mississippi to make its own way to the gulf. The voyage to Donaldsonville, about seventy-eight river miles from New Orleans, became one of the standard speed measurements for Mississippi steamers. The record — four hours and twenty-seven minutes — was set by the steamboat *Ruth*, which met an unseemly end when in 1868, some twelve miles above Vicksburg, it caught fire and burned.

The site of a trading post as early as 1750 and of a Catholic church by 1772, the town was laid out by William Donaldson, who had acquired a large tract of land there in 1806. The new town soon became known to the area's French population as *La Ville de Donaldson*. Situated as it was in the heart of sugar-cane country, it became an important shipping point for cane growers, who made of it a prosperous community of elegant homes and other attractive buildings. For three months in 1830 Donaldsonville served as the capital of Louisiana.

The next big steamer stop above Donaldsonville was Baton Rouge, the Louisiana capital city, which Samuel Clemens saw as a veritable garden in the nineteenth century, "clothed in flowers ... like a greenhouse. The magnolia trees in the Capitol grounds were lovely and fragrant, with their dense rich foliage and huge snowball blossoms."[6] The nineteenth-century capitol, built to resemble a European castle, was one of the city's chief tourist attractions. The environs of Baton Rouge presented to steamboat travelers scenes of sugar cane plantations, with elegant plantation houses, sprawling fields of cane and clusters of slave houses.

After Baton Rouge the next significant stop was Bayou Sara, Louisiana, at the mouth of the stream named Bayou Sara, just below St. Francisville, on the east side of the Mississippi. Bayou Sara, the town, had been a popular port and safe haven for flatboats since the late 1700s, and by the 1860s, with the coming of steamboats, it had grown into one of the major shipping points between New Orleans and Natchez, made so by the nearby cotton plantations that it served. Repeated flooding, however, eventually forced the town's residents and businessmen to move their homes and buildings to the higher ground of St. Francisville, situated on a bluff. During the years of the area's booming cotton economy in the mid–nineteenth century, St. Francisville became an affluent community, known for its handsome plantation homes and town houses. The town of Bayou Sara, meanwhile, declined and by the end of the century had disappeared, all but one of its buildings having been dismantled, demolished or carried away by the Mississippi's raging floodwaters.

The bluff overlooking the Mississippi at Natchez. The British novelist Frances Trollope traveled down the Mississippi in 1827 and in her travelogue wrote that Natchez appeared "like an oasis in the desert." By the mid–nineteenth century its natural beauty had become enhanced by the dozens of elegant mansions built by multi-millionaire cotton planters (Library of Congress).

Natchez, a hundred river miles above St. Francisville, was one of the few places that Frances Trollope, the early-nineteenth-century British novelist, found to her liking as she traveled down the Mississippi in 1827. "At one or two points the wearisome level line is relieved by bluffs, as they call the short intervals of high ground," she wrote in her travelogue and commentary, *Domestic Manners of the Americans.* "The town of Natchez is beautifully situated on one of those high spots. The contrast that its bright green hill forms with the dismal line of black forest that stretches on every side, the abundant growth of the pawpaw, palmetto, and orange, the copious variety of sweet-scented flowers that flourish there, all make it appear like an oasis in the desert. Natchez is the furthest point to the north at which oranges ripen in the open air, or endure the winter without shelter. With the exception of this sweet spot, I thought all the little towns and villages we passed wretched-looking in the extreme." By the mid–nineteenth century it was not only Natchez's natural beauty that made it attractive but the hundreds of mansions built by wealthy cotton planters who made Natchez a city of millionaires.

Unavoidable for steamboat travelers was the most notorious part of the city—the dockside section known as Natchez-Under-the-Hill, a rude cluster of saloons, gambling joints and brothels built on the mud flats beside the river, always the first and last part of Natchez that steamboat passengers saw.

Being a major port on the river, Natchez became one of the most prominent speed-measuring destinations for steamboats operating out of New Orleans. The steamer *Ruth* held the record for that run, too, making the trip from New Orleans to Natchez, about 350 river miles, in fifteen hours and four minutes in 1867, a time unsurpassed until 1909, when the battleship USS *Mississippi* made the run in fourteen hours, after starting two miles farther up the river than had the *Ruth*.

Vicksburg, seventy-five river miles above Natchez, situated where the Yazoo flows into the Mississippi from the northeast, is another city built on a hill that rises from the riverbank. Its known history began in 1715 with a French-built fort, Fort St. Pierre, which became Fort Nogales under the Spanish administration in 1719 and was renamed Fort McHenry after the Americans took it over in 1811. The town was named for the Methodist preacher, Newitt Vick, who bought 1,100 acres atop the bluff to build a community there. By 1826, when Vicksburg was incorporated, it had become a thriving town, enlivened by the steamboat traffic that came to carry the area's cotton away. Its riverfront grew to become almost as boisterous and disreputable as Natchez-Under-the-Hill. By 1860 the town's population had increased to 4,600. (Before the nineteenth century ended, Vicksburg gained distinction as the birthplace of Coca-Cola. Joseph Augustus Biedenharm, a candy-store and soda-fountain owner, in March 1894 put his popular soda-fountain drink in bottles that he could take out and sell in the countryside, and thus was the popular soft drink born.)

From Vicksburg, steamers continued upriver to Lake Providence, Louisiana, on the west bank—so named, it was said, because its sheltered landing, beside the lake of the same name, provided refuge from river pirates in the 1700s and early 1800s. Clemens called Lake Providence "the first distinctly Southern-looking town you come to" on a voyage down the Mississippi. After Lake Providence, it was on to Greenville, Mississippi, on the east bank, another prominent cotton shipping point.

For many years Napoleon, Arkansas, at the mouth of the Arkansas River, was a major port on the river, the next one above Greenville. There where Marquette and Joliet had halted their exploration of the lower Mississippi and two Indian villages had welcomed them in 1673, there eventually rose a European settlement that by 1832 was large enough to warrant a post office. In 1851 the town was visited by Peter Daniel, an associate justice of the United States Supreme Court, who wrote about it in a letter to his daughter, telling her, "I reached this dilapidated and most wretched of wretched places at noon today and am compelled to wait until 2 P.M. tomorrow for the mail boat to Little Rock. This miserable place consists of a few slightly built, wood houses, and the best hotel in the place is an old, dismantled steamboat."[7]

Nevertheless, Napoleon became a thriving community, its prosperity owing to the cotton crops of plantations in the area. At its peak, Napoleon had a population estimated at 2,000, plus a large but uncounted number of transients. It was the county seat of Desha County until 1874, when the county seat was moved to Watson after the river ate away a section of the riverbank and a number of buildings were washed away in the powerful flow of the river. That event marked the beginning of the end for Napoleon. By the 1880s there was nothing left of it.

Upriver from the Napoleon site is Helena. In the mid- and late 1800s Helena was the second largest city in Arkansas, with a population of about 5,000. More than merely a cotton center, the city prospered from its commerce in lumber and grain, and it was home to a foundry, machine shops, mills and wagon factories, all of which made it a major stop for steamers.

Memphis was next. The Spanish explorer Hernando de Soto is believed to have been the first European on the site of Memphis, having arrived there in the 1540s. By the 1680s French explorers had erected Fort Prudhomme there, and by 1796, when Tennessee was admitted to the Union, the site was occupied by the new state's westernmost settlement. The community was established as a town in 1819 by General Andrew Jackson, Judge John Overton and General James Winchester on a 5,000-acre land grant. In 1826 it was incorporated as a city, named for the ancient Egyptian metropolis on the Nile. It became a hugely prosperous cotton trading center, where more than 40 percent of the nation's cotton crop was traded, making of its waterfront a bustling shipping point thronged by steamboats. Prosperity swelled the city's population from 1,800 in 1840 to more than 18,000 in 1860.

After Memphis came New Madrid, Missouri, on the west bank. First established in 1789 and nearly destroyed in the 1811 earthquake that made it famous, the town, situated between the river and the forest, had been rebuilt and repopulated and had resumed its position as a regular stop for riverboats throughout the nineteenth century.

Hickman, Kentucky, one of the next small stops, was noticeable for its warehouses that held the region's tobacco crop till it could be shipped out aboard steamers.

Then came Cairo, at the extreme southwest tip of Illinois, where the Ohio River delivers itself into the waters of the Mississippi, demarcating the lower Mississippi from the upper Mississippi, some one thousand miles above New Orleans. Protected by levees, the town stands on a narrow peninsula created by the two rivers as they rush toward their confluence. Because of its strategic position at the mouth of the Ohio, the site was a natural for some sort of settlement and fortification, as the Jesuit priest and explorer Pierre Francois Xavier observed in 1721. The settlement that resulted was first incorporated as a city in 1818, and after faltering in its development — for the first few decades of the nineteenth century it had only two buildings, one a log cabin and the other a warehouse — the community made a new start in 1837 and in 1858 was re-incorporated. By 1860 it had become an important steamboat port, and its population had risen to more than 2,000.

At Cape Girardeau, Missouri, the next stop, a hill rises quickly from the riverbank, and the city is built on that hill, as if holding its feet out of the water. In the early 1800s about a dozen families comprised the community, but in the mid– and late 1800s visitors arriving by steamboat could see the Jesuit school for boys that had been built not far up the hill and, above a sloping lawn, the public college that stood farther uphill, two institutions that helped account for Cape Girardeau's reputation as the Athens of Missouri. The town had begun about 1793 and by the end of the nineteenth century it had become the busiest port on the river between St. Louis and Memphis.

Above Cape Girardeau, conspicuously standing out from the wooded hills, is a natural feature that helps vary the scenery, a sixty-foot-high rock called Grand Tower. It's about an acre in area and rises from the river near the Missouri side. The town of Grand Tower, on the Illinois shore, opposite the rock, was another steamboat stop, once known as Jenkin's Landing.

Ste. Genevieve, the next stop, believed to be the oldest European settlement in Missouri, was another town populated by no more than a dozen or so families in the early 1800s. Steamboat passengers arriving from the lower Mississippi and disembarking at Ste. Genevieve may have been surprised to discover that many of the town's structures were built of logs standing vertically on the ground, French style, with no foundation, or on a sill, rather than logs laid horizontally, one on top of the other, the usual American way of erecting log buildings. Three of Ste. Genevieve's so-called *poteaux en terre* (posts in the ground) structures have survived into the twenty-first century.

The 200-mile voyage between Cairo and St. Louis offered scenery that differed noticeably from what steamboat passengers could see on the riverbanks below Cairo — hills lush with green foliage bordering the river on both sides, a welcome relief to the lower Mississippi valley's extensive flatlands.

St. Louis riverfront around 1870. From a French trading post in 1764 St. Louis blossomed into a booming American metropolis by the 1850s, when steamboat commerce made it the largest city west of Pittsburgh and steamboats stretched for a mile along its busy wharves (Library of Congress).

Passengers traveling upriver past Cairo knew they had entered a new phase of their journey.

When they reached St. Louis, mid–nineteenth-century travelers could see what a vibrant boom town it was, its riverfront alive with steamboats taking on and discharging passengers and freight. In the 1850s St. Louis was the largest city west of Pittsburgh, its population swelled by an immigration influx in the 1840s that brought thousands of Germans, Italians and Irish to the city. From fewer than 20,000 residents in 1840 St. Louis grew to almost 78,000 in 1850 and to more than 160,000 by 1860, despite a cholera epidemic in 1849 that took the lives of nearly 10 percent of the city's population. In addition to its permanent residents, St. Louis in the mid–1800s had its full share of transients — steamboaters stopping over on their way to or from elsewhere.

The city had begun as a trading post in 1764, built on the site of ancient Indian mounds by Pierre Laclede and his teen-age stepson, Auguste Chouteau, who with a small group of men had managed to make their way up the Mississippi from New Orleans. There near the mouth of the Missouri River, the avenue to the wild and wide-open West, the little settlement quickly became a fur-trading center and drew scores of new residents. Like New Orleans, it passed from France to Spain and back to France, then to the United States as part of the Louisiana Purchase in 1803. By then St. Louis's population had increased to about a thousand residents.

In May 1804 St. Louis was the jumping-off place for Meriwether Lewis and William Clark on their historic exploratory journey into the vastness of the West, and it was to St. Louis that the two explorers returned with a wealth of new knowledge in September 1806. St. Louis was incorporated as a town in 1809, and as a city in 1822, following Missouri's admission to the Union as a state in 1820.

It was the coming of the steamboat, though, that turned St. Louis into a booming metropolis. The *Zebulon M. Pike* was the first steamer to ascend the Mississippi all the way to St. Louis and when it landed at the St. Louis riverfront on July 27, 1817, it became the first of many hundreds that would dock there. In the mid–1800s river travelers arriving at St. Louis could watch as their boat took a place in the mile-long row of steamboats that crowded the bustling riverfront.

Now, in the summer of 1870, two other steamers, two of the century's swiftest and best, were rushing toward St. Louis to see which of them would be first to join the crowd of steamboats gathered at that riverfront.

· 3 ·

The Early Going

From the after end of the *Robert E. Lee*'s hurricane deck, Captain John Cannon, standing with his friend Doctor Smyth and the Louisiana governor, could see across the boat's stern and above the turn of the river the towering columns of black smoke rising from the *Natchez*'s chimneys, about a mile behind the *Lee*. Pounding and splashing toward them, the *Natchez* was slowly but steadily closing on the *Lee*, having gained a minute on it after eight minutes into the race, despite having to plow through the rough water of the *Lee*'s turbulent wake. Gamblers taking bets from onlookers along the riverbank began to lower the odds given on the *Robert E. Lee*.

Cannon and his vessel were making good time, though. They reached Carrollton, at the western tip of the giant curve the river makes at New Orleans, about eight river miles above St. Mary's Market, in twenty-seven and a half minutes.

Some of the race's observers, standing with their watches on the levee at Carrollton, reported the time difference between the *Lee* and the *Natchez* at three and a half minutes, others as much as four minutes. At Thirteen Mile Point, the difference was reported variously at three and a half minutes, three minutes and fifty-four seconds and four minutes. By the time the two steamers passed Twenty-two Mile Point the difference was reported to be four to four and a half minutes. At that point the streaking *Robert E. Lee* had wiped out whatever gain the *Natchez* had made and was lengthening its lead. Captain Cannon and his friends were breathing easier.

Giving his account of the early hours of the race, the *Picayune*'s reporter wrote, "The purser of the *B.L. Hodge* informs us that he met the steamers thirty miles from the city, and that it took the *Hodge* four minutes to run from the *Lee* to the *Natchez*. The *Hodge* was believed to be going about fifteen miles per hour, and the racing steamers about eighteen miles." Hopeless of computing the time difference between the racers himself, the reporter allowed,

The *Robert E. Lee* takes the lead. Observers watching the race from the levee at Carrollton, some eight river miles above the official starting point, reported the *Lee*'s lead to be as much as four minutes over the *Natchez* (Library of Congress).

"Those of a mathematical turn of mind may figure out the difference of time for themselves."[1]

Then, also about thirty miles out of New Orleans, an emergency struck the speeding *Robert E. Lee*. The news came to Cannon from his chief engineer, William Perkins, and he immediately left Governor Warmoth and Doctor Smyth and rushed down to the main deck and then into the hold to reach the scene of the problem. Through the hold, empty of freight, Cannon followed Perkins and Tom Berry, the first assistant engineer, to where a five-inch pipe that fed heated river water into the *Lee*'s eight boilers had come apart at a joint, shaken loose by the vibration of the boat's huge, pounding engines. The vibration was so severe, according to the *St. Louis Republican* reporter traveling aboard the *Lee*, that he found it difficult to write with a pencil on paper as he attempted to complete a dispatch to his newspaper. "At her highest speed," he remarked, "they [the engines] cause such a vibration that it is almost impossible to write on the tables of her saloon."[2]

Hastily the engineers, their hands protected with heavy gloves, forced the separated ends of the pipe back together, inserted packing material into the joint and tightened the sleeve over it with their wrenches, reducing the escaping flow to a seeping trickle that could be tolerated. Perkins pronounced it good enough. A perfect repair would require the *Lee* to stop and shut down its machinery while the repair was made. The race precluded that. Instead,

Perkins ordered two crewmen to station themselves beside the pipe and tighten the connection whenever the engine vibration threatened to separate it again. He also ordered the bilge pumps kept running to remove the leaked water from the hold.[3]

According to one account, Perkins two weeks earlier had advised Cannon to put the *Robert E. Lee* in the drydock at Mound City, Illinois, and have its machinery undergo maintenance, but Cannon had put him off and delayed taking his advice, apparently at that time not intending to agree to the race.

Now relieved that a crisis had been averted, Cannon climbed back up to the hurricane deck to rejoin Warmoth and Smyth, smiling reassuringly into their anxious faces as he strode toward them. No sooner had he resumed his conversation with the two men than the boat's carpenter, John Buist, approached him to report another problem. The boat was too rigid, Buist announced, and the hog chains needed to be loosened to allow the boat to sag a little and lie as flat as possible in the water to offer the least possible resistance to the oncoming stream. (Hog chains were wrought-iron rods that extended from one end of the hull to the other, creating stiff braces that prevented the hull from hogging — that is, arching up like a hog's back, thus giving the device its name — and from sagging. The effects of using hog chains were to allow hulls to be built longer, shallower and with lighter timbers, thereby increasing the boats' payloads. They had been introduced into steamboat building sometime between 1835 and 1841 and were considered a major technological improvement.) Cannon strode off with Buist, but apparently did not agree to the adjustment of the hog chains.

Scores of spectators still stood waving and shouting from the east and west riverbanks, now a mile apart, as the steamers continued their march, the sun slipping below the trees on the west side of the river. As darkness came on, bonfires sprang up along the levees, providing a lighted path for the boats to steer upon as they sped into the evening, and spectators fired guns and cheered as the *Robert E. Lee* passed them.

When Warmoth and Smyth decided it was bedtime, they made their way down from the darkened hurricane deck to the cabin and found the stateroom they would share. And there was more than the stateroom to share. All furniture except one double bed had been removed from the room when Cannon ordered the *Lee* stripped, and the two men would have to sleep together in that one bed. They were political allies and close friends, though, and apparently didn't mind. Warmoth took the occasion to point out to Smyth, an Irish immigrant, the wondrous opportunities of America. "If you had not come to this country," he told him, "but had remained in Ireland, it would have been a long time before you could have slept with a governor."

Quick to reply, Smyth in his brogue gave Warmoth a not-too-subtle gibe for his carpetbagger's advantage. "If we both had lived in Ireland," Smyth said, "it would have been a damned sight longer time before you would have been a governor!"[4] It was all in fun, though.

Other passengers lingered for a while on the promenade of the deck and in the grand saloon, then they, too, ambled off to their staterooms for the night. Cannon, though, kept his post on the hurricane deck, his eyes often turning aft, where in the distance on straight stretches of the river he could occasionally see the glow of the *Natchez*'s fires when the furnace doors were opened.

The *Lee* reached Donaldsonville, about two-thirds the distance from New Orleans to Baton Rouge, after four hours and fifty minutes. The *Natchez* reached Donaldsonville after five hours and five minutes. The *Lee*'s lead had increased to fifteen minutes. Around midnight the *Lee* passed Conrad's Point, at the beginning of a long, straight stretch below Baton Rouge, and was nine miles ahead of the *Natchez*, still lengthening its lead.

Shortly before midnight more bad news came to Cannon, and he hurried down from the hurricane deck to see for himself the latest thing to go wrong. One of the boat's eight boilers had sprung a leak, and water was escaping from it faster than the boat's intake pump could replenish it. The gauges showed the steam pressure dropping, and as the water level in the leaking boiler fell, the danger of an explosion fearfully rose.

Doubtless remembering the disastrous explosion of the *Louisiana*'s boilers, Cannon ordered the furnace doors opened and he stared into fiery chambers to see if he could spot where the leaking water was entering the furnace from the boiler above it. The site of the leak would have to be found quickly and once it was discovered, repairs would have to be made immediately. If the leak were not stopped, water would drain away from the leaking boiler and the boiler's quarter-inch-thick iron plates would become overheated and burst apart, their fragments smashing into and bursting the other boilers, which would erupt in a devastating explosion and spread the furnace fires to the entire vessel.

Ordinarily in such an emergency, the vessel would immediately put into shore and tie up to a tree while the furnace fires were doused with water and the boilers allowed to cool before a crewman would crawl into the boiler to find the leak and patch it. That procedure would take hours, perhaps days. The *Robert E. Lee* would be sitting idle while the *Natchez* steamed triumphantly past it and on to victory at St. Louis. Captain Cannon hated that thought. From the open furnace door, he and his engineers could see a spot where the furnace fire had been extinguished and where steam rose from the bed of the furnace and they guessed that the leak was from a connection just above that point. The connection was from the No. 4 boiler, which must be where the leak was, they reasoned. Somebody would have to crawl under the No. 4 boiler to find the exact spot and devise a way to patch it. In the meantime, the slowed *Robert E. Lee* would continue its course toward St. Louis. And all concerned would hope for the best.

Chief engineer Perkins, though willing, was too old and not nearly limber or dexterous enough to squeeze himself into the hot, cramped, two-foot-high crawl space beneath the furnace and boilers to find the leak and make the repair. Tom Berry, the first assistant engineer, was too large a man to fit into the space, and while four other assistants stood deciding whether to volunteer, assistant engineer John Wiest stepped up and said he would try it.

The leak was found to be actually in the mud drum, the long, cylindrical, troughlike device below the boiler and connected to it. Its function was to collect the sediment that was suspended in the water that was pumped from the river into boilers. The No. 4 boiler was the fourth boiler from the right side of the row of eight boilers. Its position between the other boilers made the underside of it probably the hottest spot beneath the boilers. In an attempt to cool off the bottoms of the boilers as best they could, crewmen brought out one of the boat's hoses and sent a stream of water onto the bellies of the boilers.

Wiest stripped off his clothes and put on a set of heavy, protective overalls. He tied a handkerchief around his head to protect it from the heat and pulled on a pair of thick gloves to protect his hands. Equipping himself with a hammer, a cold chisel and a poker borrowed from the firemen who manned the furnaces, he lowered himself to the deck and on his stomach wriggled his way into the space beneath the No. 4 boiler's mud drum, then twisted his body to lie on his back. Not knowing whether he would be scalded or suffocated, he braved the searing heat and with the cold chisel in one hand and the heavy hammer in the other, he pounded away at the rivets until he had forced out enough of them to pry back a section of the iron plating. Through the opening he had created he was able to stab with the poker and widen a hole in the tile bed of the glowing furnace, close to the mud drum. Peering through it, he could see that it was not one leak but many and they were in the top flange of the mud drum, where the No. 4 boiler connected to it. Water was spraying from a number of small, rusty perforations.

Then suddenly he blacked out, overcome by the stifling heat. Cannon and two other steamboat captains, anxiously watching him, saw his body go limp, and immediately crawled into the space, grabbed Wiest by the legs and hauled him out, then placed him on the starboard guard, the extension of the

The Currier & Ives imaginative depiction of the race. The *Robert E. Lee*'s lead was threatened, although not to the extent shown in the Currier & Ives print, when one of its boilers sprang a leak below Baton Rouge and Captain Cannon had to reduce the boat's speed while dangerous, makeshift repairs were made (Library of Congress).

deck that projected past the boat's hull. In the fresh night air, Wiest soon regained consciousness and when his head cleared, he reported what he had discovered in the one good glance that he had got.

One of the group that was gathered around him suggested putting small pieces of hemp, a little at a time, into the No. 4 boiler's water line — a trick that probably had been performed in engine rooms before. The bits of hemp packing, suspended in the water that was leaking through the perforations, would lodge themselves in the holes and stop them up. All agreed it was worth a try.

The engineers forced the fragments of hemp packing, which they chopped into small bits, into the intake suction valve, then restarted the intake pump, sending the hemp fibers coursing through the water line. They switched off the pump and inserted more hemp into the line and again started the pump, fixing their eyes on the gauges. After several applications of hemp into the water line, the gauges finally showed that the pressure in the boiler had stopped falling and had gradually begun rising. The hemp fibers had become minuscule fingers in the dike.[5]

The *Lee* was now just above Plaquemine, Louisiana, a community on the west bank of the river, and was steaming for Baton Rouge. Although the hour was past midnight, excited spectators had climbed into skiffs and put out from the riverbank at Plaquemine, battling the *Lee*'s wake to hail and cheer the grand steamer and its crew. They would soon also be cheering the *Natchez*, which could be seen from the stern of the *Lee* by the glow of its furnaces when their doors were opened. It was only about four hundred yards behind the *Lee*, which was increasing its speed to regain the time it had lost during the latest emergency.

Beset by worry over the condition of his vessel, Captain Cannon was having doubts about continuing the race. Still on the hurricane deck, he called his old friend John Smoker over to him and asked him what he thought about ending the race at Baton Rouge and declaring the *Robert E. Lee* the winner to that point. Smoker didn't think much of the idea. "As long as we're ahead," he responded, "we'd better keep so." Thus encouraged, Cannon gave up the thought of stopping, for the present anyway, although he continued to worry.

With the *Natchez* hot on its heels, the *Lee* passed Baton Rouge, on the east bank of the river, about three o'clock in the morning on Friday, July 1. Beneath the lights on the wharf clumps of bleary-eyed spectators watched as the two boats steamed past, first the *Robert E. Lee* and minutes later the pursuing *Natchez*. By the time the *Lee* reached Bayou Sara, just above Baton Rouge, it was ten minutes ahead of the *Natchez*, having made it that far in ten hours and twenty-six minutes.

The reporter from the *St. Louis Republican* on board the *Robert E. Lee* was as wide awake as its captain, recording the events of the race and the passing scenes observed from the vessel's decks:

> The scene from time of departure till dark ... baffles description. As we steamed along the watery race track, the whole country on both sides of the river seemed alive with a strange excitement expressed in a variety of gestures, the waving of handkerchiefs, hats, running along the river shore as if to encourage the panting steamer, and now and then far off shouts come cheeringly over the waters, and were plainly heard above the roaring of the fires, the clatter of machinery, the dashing of the waters and the rushing of steam. All the life in the vicinity of the river appeared to be thoroughly aroused into the unusual activity by this struggle of two steamboats for the palm of speed. The settlements and plantations along the coast as we passed turned out their whole forces, and seemed to have taken a holiday in honor of our flying trip.
>
> Up to and beyond Plaquemine men and boys in skiffs came out almost in our track to hail us with warm welcome and get a word, if possible, with one of the officers or crew. This is but a moment. They are struck by the swells and dashed and rocked away off towards the shore, far in our wake. As long as they are in sight they wave us adieu. The inhabitants all appear to live out of doors, or are crowded in the windows or on the housetops as we approach. The most lively interest is depicted in every countenance and is uttered in every voice.

At Baton Rouge, which we reached about 1 o'clock, this morning, there were still people on the wharf, but silence had nearly been restored on shore, and during the rest of the night nothing was to be noted but the still, anxious groups on board.[6]

By the time the sun had risen on the new day, the *Robert E. Lee* was pulling farther ahead, and its unsleeping captain, unable to shake his worry over the boat's machinery, went down into the engine room and asked Perkins to slow down, telling him they were well in front of the *Natchez*, that there was no need to run at full speed. Perkins replied that it wasn't the *Natchez* he was thinking of. Rather it was the speed record for a trip from New Orleans to Natchez, which had been set by Leathers's fourth *Princess* in 1856 and which he intended to beat. The *Lee*'s chief engineer was not concerned about the boat's performance so far, and Cannon, apparently reassured, returned to his post on the upper deck.

On board the *Natchez*, another reporter from the *St. Louis Republican*, offering a different perspective of the race, observed that "The captain [Tom Leathers] is sleepless on deck, the pilots are nervous yet confident at the wheel, the engineers stand by their engines watching every movement of the machinery, and the firemen work like Trojans, and look like demons in the red glare of the furnaces."[7] The anonymous reporter took time to notice the spectacle of the steamer racing through the darkness, "cleaving the river wide open," as he put it. "The effect at night is simply grand," he wrote. "The steamer plows on her watery way, puffing white clouds and streaming a constant current of fiery sparks from her chimney tops, bounded by blackness on either side. But the people on shore are sleepless, too, and send their greetings through the darkness as we pass."[8]

Leathers had been given a gold pocket watch as a trophy for his record-breaking run from New Orleans to St. Louis less than two weeks earlier, when he had made the trip in three days, twenty-two hours and forty-five minutes, mere minutes faster than the old *J.M. White* had made it twenty-six years earlier. On that occasion, addressing an audience of well-wishers, Leathers had proclaimed with satisfaction, "Gentlemen, none of we older men will live to see this time beaten, and probably few of the younger ones."[9] Now at about eleven-thirty P.M., standing on the boiler deck, staring over the *Natchez*'s bow, straining to see if the distance between the two boats was closing, he checked the watch as the *Natchez* passed the one-hundred-mile point upriver from New Orleans and concluded that the *Lee*, which had passed the hundred-mile point six minutes earlier, had gained no more than half a mile on the *Natchez* after running for a hundred miles. Leathers's boat had not reduced the *Lee*'s lead, but it had not let the gap substantially widen either.

Leathers checked his watch again as the *Natchez* reached Plaquemine, one hundred and thirteen miles from New Orleans. Making good speed, his boat then had covered thirteen miles, from the one-hundred-mile point, in forty-five minutes. Yet, as it raced toward Baton Rouge, it had not closed on the *Robert E. Lee*. At Baton Rouge Leathers looked at his watch once more. Eight hours and twenty-eight minutes had elapsed since he had passed St. Mary's Market. And the *Lee* was still ten minutes ahead of him. The *St. Louis Republican*'s reporter, observing the *Natchez*'s captain, described the scene and the mood:

> There was not much conversation. Capt. Leathers remained but a short time on the roof and then sat on the boiler deck absorbed in thought. The engineers watched carefully every movement, the firemen worked like Trojans, and looked like demons in the red glare of the furnaces....
>
> Heavy swells from the Lee are still striking the shores, and, to confess it, impeding our progress. But the Natchez still plows on her way, puffing white clouds and streaming a myriad of sparks from her chimneys. A wide breadth of the river is lighted up in front of the boat.[10]

Beyond Baton Rouge, the *Lee*'s lead diminished slightly. When the boats reached the mouth of the Red River, the next checkpoint above Bayou Sara, their traveling times from St. Mary's Market were identical — twelve hours and fifty-six minutes — although the *Lee*, having started ahead of the hard-charging *Natchez*, remained in front. By the time it reached Stamps's Landing, upstream of the Red River's mouth, the swift *Robert E. Lee* had increased its lead again, gaining four minutes on the *Natchez*. It was two miles ahead. Two checkpoints later, at Briar's Landing, the speeding *Lee* had widened its lead still more.

It was mid-morning on Friday when from the decks of the *Lee* the city of Natchez was sighted, standing atop the high bluff that rose from the river's edge. It was at the Natchez waterfront that the six-prong gilded antlers, mounted on a carved deer's head, were kept, the speed trophy awarded to the steamboat that made the fastest time between New Orleans and Natchez, then estimated at 296 river miles. Leathers had won the antlers — the horns, as they were called — for his record-setting run fourteen years earlier. He had made the trip in seventeen hours and thirty minutes, a time that many along the river believed impossible to beat. The *Robert E. Lee* was about to prove them wrong. Aboard the *Lee*, the St. Louis reporter wrote: "In a few minutes we will be opposite Natchez. The morning is beautiful, and everything is lovely."[11]

The *Lee*, slowing down to take on fuel, approached the levee, "thronged all morning," as one reporter wrote, "by immense and enthusiastic crowds of all colors and conditions," and came abreast of the Natchez wharf just about 10:15 A.M. The band that had been assembled at the waterfront, expecting to hail the city's favorite in the race, the *Natchez*, as it glided in first, was so dismayed by the *Lee*'s arrival ahead of the *Natchez*, that it refused to play a note for Cannon and his boat. The spectators massed at the river's edge were more sportsmanlike, though, breaking out into loud cheers. "Great crowds on the wharf," the St. Louis newsman aboard the *Lee* reported, "and when we left, the wildest shouts went up. Every heart on board was touched with excitement. The tension of the nerves is continual and almost painful at times. Truly the *Lee* is a thing of life."[12]

The *Robert E. Lee* had made Natchez in seventeen hours and eleven minutes, beating the *Princess*'s record time by nineteen minutes. Bettors who had put their money on the *Lee* in this first important phase of the race were exultant. In Natchez, though, "betting immediately fell to zero," the St. Louis reporter observed, "everybody wanting odds on the *Natchez*."[13]

The *Lee* slid by the wharf boat, a floating, covered dock moored in the river, without stopping, and the *Lee*'s Natchez agent, responding to shouts of "Take down those horns!" coming from passengers and crewmen aboard the *Lee*, jumped aboard with the horns, prettily trimmed with flowers and ribbons. The antlers made a handsome trophy, attached to a polished wood plaque with an inscription, apparently dictated by Leathers, that read: "Why Don't You Take The Horns? Princess' Time To Natchez, 17 Hours and 30 Minutes." Cannon took the coveted horns and blew the *Robert E. Lee*'s steam whistle in acknowledgment.

But he didn't land. The *Lee* merely slipped up between two barges loaded with sacks of coal and had the barges tied to the *Lee* as the sacks were unloaded onto the *Lee* while it continued upstream, all done by a prior arrangement made by the *Robert E. Lee*'s canny captain. Once the coal was aboard the *Lee*, the barges were cut loose and allowed to drift back to the Natchez waterfront.

By the time the *Natchez* arrived, eight minutes behind the *Lee*, and tied up to the wharf boat, Cannon and the *Lee* were steaming away in the distance, leaving the *Natchez* and its disappointed fans, some of them openly weeping, far behind. The *Natchez* did manage to beat the old record of the *Princess*, by eleven minutes, reaching the Natchez shore in seventeen hours and nineteen minutes from the starting point at St. Mary's Market. But its time was not good enough to prevent the horns from passing into Cannon's hands. And the trailing *Natchez* lost another eight minutes as it put ashore twelve dejected Natchez-bound passengers and took on fuel.

Leathers then rushed the *Natchez* back out into mid-river to resume the race. Try as desperately as he might, though, to close the distance between him and the *Robert E. Lee*, he was finding that the *Natchez*, with its thirty-four-inch cylinders, lacked sufficient drive to outrun the *Lee*, powered by forty-inch cylinders, on the river's long straightaway.

News of the *Robert E. Lee*'s record-breaking feat quickly spread by telegraph to New Orleans and elsewhere and was made public. The *Picayune* reported, "At an early hour ... a large crowd of eager persons gathered around the Picayune Office to hear the news, and all over the city the most intense interest was manifested. What with the shouting of the news boys — each of whom had something staked on the result — and the cheering whenever a new dispatch from up the river came in, one was forcibly reminded of the war times just after some tremendous engagement."[14]

Leathers, outraced to Natchez, now conceded that he had misjudged the *Lee*'s ability. "I've underestimated her power," he confessed, but then seemed to grow more determined than ever. Above Natchez the Mississippi narrowed and became more twisting. The *Lee* may have had an advantage on the broad, straight sections of the river below Natchez, but the sleeker, more maneuverable *Natchez*, Leathers believed, would have the advantage now in channels studded with small islands and sand bars and salients thrusting out from the banks. Besides, the *Natchez*, he was certain, had the two best pilots on the Mississippi, Frank Clayton and Mort Burnham. The race now would be not simply a contest of speed but of piloting skill. It was far from over.

Up ahead were the towns of Cole's Creek, Waterproof, Rodney and St. Joseph, spotting the riverbanks in Louisiana and Mississippi. St. Joseph, Mississippi, a frequent stop, was the home of Ed Snodgrass, a merchant who held the distinction of being a friend of both Cannon and Leathers, and loving a bit of mischief, he delighted in passing on to each the insults of the other when their boats tied up at the St. Joseph landing. Leathers had recently had a painful, bothersome carbuncle develop on his back, suffering so severely that he brought his doctor along with him on the *Natchez*. Seeing him in his torment, Snodgrass had sympathized, but after the *Natchez* departed and Cannon arrived on the *Robert E. Lee*, he eagerly informed Cannon of Leathers's condition.

"A carbuncle, huh?" Cannon responded.

"Yes," Snodgrass answered.

"Well," Cannon said, "you tell the old *blankety-blank-blank* that I had a brother — a bigger, stronger man than I am — and he had one of them things and died in two weeks."

When Cannon took a misstep aboard the *Robert E. Lee* one time, he fell to the deck and broke his collarbone. That news reached Snodgrass and was passed along to Leathers, who instructed Snodgrass to tell Cannon, "I wish it had been his *blankety-blank* neck."[15]

After passing St. Joseph, despite its best efforts to catch up, the *Natchez* was still more than eight minutes behind the *Robert E. Lee*. To make matters worse, it had to make a stop at Grand Gulf, Mississippi, while the *Lee* would be able to keep up its steady pace. At Grand Gulf, which it reached a little past 5:15 P.M. on Friday, the *Natchez* took on ten passengers who were Captain Leathers's regular customers and who were making their annual trip north. They had booked passage on the *Natchez* on its previous trip, and Leathers, true to his word and the notice he had placed in the *Picayune*, was taking care of his customers. But he lost another eight minutes in getting the passengers and their trunks and other baggage aboard, which could only have further distressed him, having learned that the *Lee* had steamed past the landing twenty minutes earlier.

For all his braggadocio, gruffness and intimidating manner, Tom Leathers had a heart that could be touched, and his faithfulness to his Grand Gulf passengers was but one example. His generosity showed in the number of times he had given free passage to ministers, priests and nuns and to individuals who were desperate for transportation but unable to pay. Sometimes he even put them in staterooms aboard the *Natchez*. The professional gambler George Devol, a frequent passenger on Leathers's boats, once came upon a woman who needed passage for herself and her six children but was unable to pay the price of the tickets. Moved by the woman's plight, Devol doffed his top hat and passed it among the *Natchez*'s passengers and officers while the boat was docked. According to Devol, all of them put something in the hat except for one man. Devol then took the hatful of bank notes and silver coins to Leathers, standing on the hurricane deck, showed it to him, told him about the poor woman and said what he had collected should be enough to pay for tickets for the woman and her kids. Leathers, though, refused to take the money. "Give the money to the woman," he told Devol. He then instructed the *Natchez*'s chief clerk, Samuel Ayles, to book the family into a stateroom and treat them as if they had paid the full first-class fare.

Devol made his way back to the woman, gave her the money he had collected and returned to the *Natchez*'s saloon, where he took over a table and opened up a game of three-card monte. One of the first players he attracted was the man who had declined to contribute when Devol passed his hat. Devol took him for eight hundred dollars, to the delight of the other passengers, who taunted him, one of them asking, "Aren't you sorry you didn't give something to the woman before you lost your money?"

The man complained to Leathers, to no avail. Leathers refused him both help and sympathy.[16]

After Grand Gulf came Hard Times, Louisiana, and then Vicksburg would be next. The St. Louis reporter on the *Natchez* narrated the voyage:

> The scenes on board, as we witness the crowds and hear the shouting, cannot be portrayed. At this hour we are approaching Vicksburg, the Lee being still considerably ahead. But we are surely, though slowly lessening the distance.
>
> Sometimes in a long stretch of clear river she is plainly in sight, then a bend shuts her out, all but her smoke, which hangs away off northward like a dense cloud; then an island or a sudden projection of woodland hides all traces of our lively rival from our view. We feel safe but keep wonderfully busy, because we know she is there going like lightning. There is life and wakefulness and speed and determination in the swiftly following vessel, which will give us the victory before we are done with her. These occasional glimpses of the Lee seem to give the Natchez more muscle and force her to her very best.[17]

The *Natchez* had to make another stop at Vicksburg, to discharge seventeen passengers and take on more coal. On reaching Vicksburg, Leathers checked his watch and marked his time from St. Mary's Market at twenty-four hours and forty-two minutes. He had gained time on the *Lee*, but was still at least eight minutes behind. Like Cannon at Natchez, Leathers had barges loaded with coal waiting for him at the Vicksburg wharf and he tied them to the *Natchez*, pulling them alongside as he swung his boat back into the current. When the coal, packed in hundred-pound sacks, had been transferred to the decks of the *Natchez*, he cut the barges loose and resumed full speed, stalwart in his confidence that he could catch up to and overtake the *Robert E. Lee*.

The *Lee* had made Vicksburg in twenty-four hours and thirty-eight minutes from St. Mary's Market and had refueled on the run as it had done at Natchez and as the *Natchez* was to do behind it. Evidently now feeling a sense of triumph, Captain Cannon himself penned the message to be telegraphed to New Orleans, giving the elapsed time from St. Mary's Market and noting that the *Lee* was "16 minutes ahead of *Natchez*."

News of the *Lee*'s continuing lead was received in New Orleans with jubilation. The *Picayune* published a special edition, an extra, to keep its readers up to date. "When the extra *Picayune* was issued," the newspaper's reporter wrote, "with the announcement of the position of the steamers at Vicksburg, the excitement, if possible, increased, and cheer on cheer went up for the 'Bobby Lee,' till it seemed as though the people were holding a grand jubilee.... The friends of the 'Natchez' still hope that she may recover her lost time, and lead at Cairo, but there can be no doubt the chances are now in favor of the 'Lee.'

"The next point from which they will be telegraphed is Helena," the *Picayune* writer continued warily, "and there is no telling what may occur before that point and Memphis are gained."[18]

PART TWO. THE ORIGINS

• 4 •

The Pioneers

By the time he reached his fifties, Robert R. Livingston had assembled an impressive resume. He was a member of the five-man committee named to draft the Declaration of Independence (the four other members being John Adams, Benjamin Franklin, Roger Sherman and Thomas Jefferson, who headed the committee). He was elected to the New York Congress in 1776 as a representative from Dutchess County and he served on the committee that wrote a constitution for the new state of New York. The New York Congress appointed him chancellor of the state, to preside over the state's Chancery Court, and on April 30, 1789, as chancellor of New York, he administered the oath of office to George Washington when Washington was inaugurated as the new nation's first president. In 1801 Livingston was appointed by President Thomas Jefferson to be the United States minister to France, the position from which he negotiated the Louisiana Purchase.

Widely considered both brilliant and learned, Livingston was elected the first president of the New York Society for the Promotion of Agriculture, Arts and Manufactures following its organization in 1791. He liked to think of himself as a practical scientist and he devoted two hundred acres of his vast estate on the east side of the Hudson River, Clermont, to agricultural experiments, which included using gypsum in the cultivation of corn, buckwheat and clover. He also conducted experiments to improve the breeds of his cattle and sheep.

In 1797 Livingston's active mind lighted on the notion that he could build—or, rather, have built—a boat that would be powered by a steam engine. It wasn't a new idea, Leonardo da Vinci having been possibly the first to imagine one and an obscure inventor named David Ramsay having received a patent for one in 1618 and another in 1630.[1] Nothing came of those ideas, though, except their survival as possibilities in the inventive minds of persons who wished to find a way, as it was put in one patent application, of "making of Shipps to saile without the assistance of Wynde or Tyde."[2]

Another early steamboat was proposed by an English physician, John Allen, who described and patented it in 1729. Another was devised in 1736 by an English clock repairman named Hulls, whose creation, it was said, looked more like a clock than a boat. In 1760 a Swiss named Genevois came up with an idea for a boat that would be propelled by a watchlike works that would be wound by the force of steam. Throughout the remainder of the eighteenth century a succession of inventors — or men who hoped to become inventors — proposed a variety of steamboat schemes, most of them coming to nought.

The existence then of workable steam engines was keeping the steamboat dream alive. The Newcomen engine, named for its inventor, English blacksmith Thomas Newcomen, had been developed in the early 1700s, with a boiler positioned directly below a cylinder that contained one large piston. Steam entered the cylinder from the boiler and drove the piston upward, and when the piston reached the top of the cylinder, water was sprayed into the cylinder to dissipate the steam and create a vacuum, causing atmospheric pressure to draw the piston down to the bottom of the cylinder again, and the cycle was then repeated, the piston's action providing continuous movement that could be applied to a number of uses, especially including pumping water out of mines. The steamboat devised and patented by the clock repairman Hulls in 1736 used a Newcomen engine to drive a paddle wheel. There is no record, however, of the boat's having been built.

In 1783 a boat powered by a two-cylinder Newcomen engine and built by a French nobleman, Claude-François-Dorothée Jouffroy d'Abbans, actually proved workable in a short run on the River Saone at Lyons, France. It thus became the first boat to go against the current under its own power, unaided by wind or wave or the muscle of man or beast. For fifteen minutes, billowing smoke and sparks from the fire beneath its boiler, the vessel moved upriver at a speed that was as fast as a man could walk, cheered on by crowds gathered along the riverbanks. Then the boat's bottom planks, taking a terrific beating from the pounding of the pistons, broke loose, opening the hull to a torrent of river water. What was more, the boiler's seams burst, sending a cloud of steam into the air and killing the engine. But despite it all, Jouffroy managed to guide the stricken vessel to the river bank, where he leaped safely to shore, satisfied that he had achieved success for his invention. The French government, however, to which he was looking for additional financing to try again, refused to accept his account of his boat's trial run, and when France exploded in revolution in 1789, Jouffroy fled the country, his experiments ended.

In 1769 a Scottish instrument maker and inventor, James Watt (for whom the unit of power, the watt, was named), patented a steam engine that made a significant technological advancement over the Newcomen engine. Watt decided that Newcomen's engine would be much more efficient if the cylinder could be kept hot, instead of being cooled by the water sprayed into it to dissipate the steam, a process that required time to reheat the cylinder so that a new puff of steam would not immediately condense upon entering the cylinder. And so Watt devised a condenser, a container separate from the cylinder but connected to it by a valve that drew the steam out of the cylinder to provide the vacuum that caused the downward stroke of the piston. He later improved that engine by closing the top of the cylinder and introducing steam into it at both ends to create a reciprocating action that allowed the engine to generate power on both the upward and downward movement of the piston. That two-stroke design made for a more efficient, more cheaply run and much more smoothly operating engine. Its superiority soon made the Watt engine the standard of steam-engine design, even as its inventor continued to improve it and adapt it to a variety of uses.

John Stevens, a New Jersey land developer and inventor who was also Robert Livingston's brother-in-law, had become deeply interested in the idea of a steamboat. Stevens knew about the steamboat-building efforts of John Fitch, the first man in America to build workable, though faulty, steamboats, which Fitch had been doing since 1785 with limited success. Stevens also knew about the experiments of James Rumsey, another American trying to build a workable steamboat. Stevens shared their passion and their handicap — the lack of sufficient funds to develop a dependable, commercially successful steam-powered boat. Stevens saw his wealthy brother-in-law as just the financial angel he needed to make his dream come true. He talked Livingston into bankrolling the construction of a steamboat in partnership with him.

Once committed to the project, Livingston quickly became an ardent student of steam power, reading every book he could find on the subject. Even so, his knowledge of engineering did not reach the level of Stevens's, and his haughtiness prevented him from accepting some of Stevens's suggestions for the construction of a test boat. In late 1797 Livingston entered a partnership with Stevens and machinist Nicholas J. Roosevelt, who operated a foundry on the Passaic River in what is now Belleville, New Jersey. Roosevelt was the man Livingston and Stevens needed to build not only the boat but the steam engine, the British government having prohibited the export of Watt engines built in Britain, forcing the Americans to build their own.

The boat design that Livingston gave Roosevelt called for a horizontal wheel mounted below the keel to provide propulsion, despite Roosevelt's objection that the horizontal wheel would be inefficient. Vertical paddle wheels mounted on the sides of the hull, Roosevelt argued, would be much better. Livingston replied that his design was based on "perfectly new principles" and as the man who was paying the bills, he demanded that Roosevelt follow the plan.

Details of the venture were finally worked out between the three men — Livingston, Stevens and Roosevelt — and construction of the boat began in April 1798. Livingston participated from a distance, frequently issuing instructions to Roosevelt by mail, repeatedly making changes, but visiting the foundry only rarely while the work went on.

Finally, in August 1798, the building of the boat and the engine was finished and the engine was installed. It was time for a trial run. On the appointed day, however, the boat proved unable to move. Roosevelt blamed Livingston's horizontal wheel while Livingston blamed Roosevelt's steam engine. It was back to the drawing board. In October 1798, with adjustments having been made, but the horizontal wheel still in place, a new test run was scheduled. This time the boat managed to move upriver for a short distance at a rate estimated by Roosevelt as "three miles [an hour] in still water" before its heavily pounding engine shook apart both the boat and its machinery.

Livingston prepared to try again, making more changes to his design, more innovations. He tried an engine that used mercury instead of steam in its cylinder, but it, like all the other new ideas he came up with, failed to work. At last he realized he was no engineer and in February 1800 he teamed up again with Stevens in a new agreement that stated they would build a new boat and share the cost of it. With some reluctance, they again included Roosevelt in the arrangement, allowing him to contribute his foundry's labor in lieu of cash.

Livingston's attention, however, was then diverted to the task given him by President Jefferson in the fall of 1801, when he was sent to France to secure from Napoleon's government the right of the United States to use New Orleans as a shipping base.

While in France on that assignment, Livingston met an ambitious young American who captured his interest. He was Robert Fulton, the son of an Irish immigrant, also named Robert Fulton, who with his wife, Mary, had settled on a farm near Lancaster, Pennsylvania, where young Robert was born on November 14, 1765, the fourth of the family's five children. Fulton had grown up in Lancaster, a manufacturing town with shops that produced tools and farm implements as well as weapons that helped General George Washington's revolutionary army wage war against the British. Fulton became acquainted with the sorts of things that could be wrought from metal and the processes for doing so.

Fulton also had a considerable talent for painting and drawing and in 1785, at age twenty, he had set himself up in a studio in Philadelphia to make a living as a painter of miniature portraits suitable for enclosure in lockets and brooches. His portrait business did well enough that in May 1786 he was able to buy property for his mother and some of his siblings to live on in western Pennsylvania. It also allowed him to move to grander quarters, and he advertised in 1786 that he had "removed from the northeast corner of Walnut and Second Streets to the west side of Front Street, one door above Pine Street, Philadelphia."

That new location placed him within a block of the Delaware River, where in the summer of 1786 John Fitch was experimenting with a forty-five-foot boat powered by a steam engine with a three-inch cylinder and propelled by two sets of paddles, one mounted on each side of the craft. There is no evidence that Fulton witnessed, much less studied, Fitch's contraption, but it drew a sizeable crowd to the riverbank, and Fulton, naturally curious, particularly about things mechanical, may well have been one of the most interested observers of Fitch's experimental steamboat.

That same year, 1786, Fulton, troubled by a persistent cough, took time off to stay at a spa in West Virginia (then Virginia) in hopes of curing his ailment. While there he met some people to whom he showed his paintings and who were so taken with his work that they advised him to go to England and study under some of the prominent artists there, particularly Benjamin West,

Robert Fulton, inventor of the first successful steamboat. The son of an Irish immigrant, he was born near Lancaster, Pennsylvania, where while growing up he became familiar with the making of machinery and metal devices (Library of Congress).

an American who had made a name for himself as a portrait painter in England. In the spring of 1787 Fulton, carrying about two hundred dollars and a letter of introduction to Benjamin West, sailed for England, arriving in late spring. Though West and his wife received him cordially, Fulton's artwork failed to impress West enough for him to take Fulton on as a student, and eventually Fulton despaired of making a living as an artist.

His big, absorbing interest then became things nautical, particularly canals, canal locks and canal boats, all pertinent concerns of that day. He conceived ideas for improving them, but failed to find sufficient financial backing to put his ideas to work. In 1794 he corresponded with the company that James Watt and Matthew Boulton had formed, inquiring about buying from the company a steam engine that could be used to power a boat. He also came up with ideas for building a submarine, as well as designs for torpedoes and mines. In England he was unable to secure the financing he needed to develop his ideas and in June 1797 he boarded a ship at Dover and sailed across the Channel to Calais and from there traveled to Paris.

In Paris he obtained lodgings at a Left Bank pension that attracted well-to-do Americans and it was there that Fulton met Ruth Barlow, the charming Bohemian wife of the odd, affluent and socially prominent poet Joel Barlow, from Connecticut. Ruth was in Paris awaiting her husband's return from Algiers, where as a representative of the United States government he was attempting to gain the release of American seamen held hostage by Barbary pirates. Ruth, a pretty, witty woman in her early forties, became attracted to the thirty-one-year-old Fulton, tall and handsome, with dark, curly hair and an engaging charm and vigor. Upon Barlow's return from Algiers, Fulton charmed him as well, and the Barlows invited Fulton to move from the pension into an apartment with them, which he did, forming a fast friendship—and with Ruth, more than friendship, with Joel's approval—that was to last many years. In late October 1800 Barlow bought a huge, handsome mansion for the three of them, at 50 Rue de Vaugirard, across from the Jardin du Luxembourg, in one of Paris's poshest neighborhoods.

It is likely that Livingston met Fulton through the Barlows, probably at a dinner party given by them in early 1802 to show off their impressive new home, although there is no known record of the meeting. In any case, while in Paris, Livingston and Fulton became acquainted and in so doing discovered each other's interest in steam-powered boats. At age fifty-five and impaired by increasing deafness, Livingston quickly developed an admiration for Fulton's relative youth and vigor, his enthusiasm and his apparent knowledgeability about the mechanics of harnessing steam power to boats. Fulton, Livingston soon began to think, might be just the person he needed to help him create a steamboat that could navigate the Hudson between New York and Albany no matter the wind, weather or current.

Fulton of course saw in Livingston a Daddy Warbucks whose money could make Fulton's dream of creating a great invention — one that would make him the fortune he sought — come happily true. Before long, the two men reached an agreement to cooperate in the building of an experimental steam-powered boat. They decided it would be best to buy an engine rather than build one, and Livingston, with his diplomatic connections, got the job of seeking permission from the British government to export — to the United States, since Britain and France were at war — one of the engines manufactured by the firm of Watt and his partner, Boulton. Livingston also assumed the burden of footing the bill for the boat's construction. Fulton's task was to come up with a workable propulsion system and hull design — the two major problems plaguing others who were trying, or had tried, to build a practicable steamboat.

Fulton's latest propulsion idea at that time was to create what was very much like a bicycle chain, to which paddle boards would be attached and which would rotate in a long oval, propelling the boat as the paddle boards moved continuously through the water, drawn by the chain. To test the idea he ordered a three-foot-long model built, the propulsion chain for which would be actuated by clock springs. He planned to observe the working of the model, then design a full-size boat based on the proportions of the model. At the time, Fulton, a lover of the good life, was staying — with Ruth Barlow — in the fashionable mountain resort town of Plombières, where the model was to be sent to him and where he would test it in a section of a stream he had prepared for the model's run. He and Livingston, still in Paris, kept in touch by mail.

The model arrived in Plombières in late May, and Fulton promptly put it to the test. On the basis of the test, he decided that the full-size boat should be ninety feet long and six feet in the beam. Ever quick with a sharp pencil, he calculated that the full-size craft could travel at a speed of eight miles an hour, carry fifty passengers and make the New York-to-Albany voyage in eighteen hours. He figured that the boat would burn a ton and a half of coal and that the cost of the coal plus the salaries of the crew would bring the expenses to no more than twelve dollars per trip, and assuming a capacity load of fifty passengers, the profit per trip would be $198. Days later he revised his figures and projected that he could make the boat two hundred feet in length and twelve feet in the beam, have it carry one hundred and twenty passengers and attain a speed of sixteen miles an hour, shortening the running time to Albany, he estimated, though unsure of the distance, to twelve hours.

He wrote to Livingston to inform him of the progress he had made in developing the proposed boat from his experiments with the model and with small deference to Livingston's position as sole financier, he demanded a report on Livingston's progress in his part of the project. "Having now finished my experiments," Fulton wrote, "I will thank you to let me know how you proceed with yours." Fulton now could clearly see grand possibilities, all of them owing, as he saw it, to his efforts, and he wrote again to Livingston, this time to propose an agreement that would have the two of them split the profits from the boat's operation fifty-fifty. He gave Livingston his justification for such a deal. "The leading principle," he wrote with candor, "...is that I estimate my time (for it is my attentions which must carry it all into effect, and my knowledge of the subject) as an equivalent to your money."[3]

Livingston had some doubts about Fulton's boat design, thinking the slender craft might prove unable to withstand the constant pounding of the engine's piston, and he balked outright at Fulton's fifty-fifty-split proposal, telling him that "the demand you make of half the profits" was "much too great a compensation for the labour and time it will cost you."

Apparently worried that, as Livingston warned, he could not be issued a U.S. patent for his so-called endless-chain propulsion system because such a device was already in use, Fulton abandoned the endless chain and substituted vertical paddle wheels to propel the proposed boat. They of course were not a new idea either. Paddle wheels turned by oxen were used by the Romans to propel boats before the first century A.D., and warships propelled by paddle wheels turned by manpower were used by the Chinese in the seventh century A.D. Roosevelt, too, had proposed vertical paddle wheels. Nevertheless, Fulton felt he could patent *his* paddle wheel and establish exclusive rights to the device he would design. "Although the wheels are not a new application," he reasoned, "yet if I combine them in such a way that a large proportion of the power of the engine acts to propel the boat in the same way as if the purchase [of the wheel] was upon the ground the combination will be better than anything that has been done up to the present and it is in fact a new discovery."[4]

After some sparring over the details of a partnership arrangement between them, Livingston, evidently seeing Fulton as his best hope for realizing his steamboat dream, at last agreed to the fifty-fifty split on profits. He did so on the condition that Fulton also invest money in the enterprise. On October 10, 1802, the two men signed an agreement to form a company that would seek a U.S. patent for "a new Mechanical combination of a boat to navigate by the power of a Steam Engine" and that would run between New York and Albany at eight miles an hour in still water and carry at least sixty passengers, allowing two hundred pounds per passenger. Livingston and Fulton each subscribed to fifty shares in the company. Fulton would go to England and build a prototype of the same specifications, using an engine to be borrowed from Boulton and Watt.

While Livingston was anteing up the capital over the next couple of months, Fulton tried to persuade Boulton and Watt to let him borrow one of their engines. His request was firmly rejected, and in the spring of 1803 he went to French machinist Jacques Perier to ask him to build the engine. Fulton had worked with Perier four years earlier, in 1799, when his bright idea of the moment was a rope-making machine, an idea that failed a year later when he ran out of money and financial backers. Perier had also collaborated with Fulton on the construction of the submarine Fulton had conceived, another scheme that failed. Perier had not only visited the Boulton and Watt foundry and machine shop in Birmingham and observed its operations but had also acquired a Watt engine and installed it in a Paris waterworks. Now he agreed to build the engine that Fulton — and Livingston — needed for their experimental boat. Etienne Calla, the man who had built the three-foot model, would put together the boiler and some parts of the engine.

This time Fulton would use vertical paddle wheels for propulsion, one mounted on each side of the boat, instead of the rotating chain he had used on the model. As for the hull, he boned up on the studies made by others, including renowned French mathematician Charles Bossut and English researcher Charles Mahon, Earl Stanhope, to determine the most efficient shape, the one likely to effect the least resistance as the craft moved through the water. Fulton further had the advantage of having seen drawings of the steamboat model designed by a French inventor named Desblancs — which had been on exhibit at the Paris Conservatory of Arts and Trades and of which Joel Barlow had made a sketch that he had sent to Fulton at Plombières. Fulton might also have seen the plans for Claude Jouffroy's boat, which were in Jacques Perier's possession, and he knew about John Fitch's experimental boats, plans for which Fitch had left in Paris. And of course he knew about the creations of Livingston, Stevens and Roosevelt, as well as about the attempts of others.

After much contemplation, Fulton, a painstakingly careful worker, finally decided on a flat-bottomed, keelless hull with a tapered bow and stern. The boat would be steered with a rudder and tiller. Its two paddle wheels would each have ten spokes with rectangular boards attached at their farthest ends to serve as paddles. The engine, held in place and supported by a special framework, would be positioned amidships, mounted on the boat's bottom planks, which would be reinforced beneath the piston and the boiler. The engine and its components would take up about half of the space in the hold.

His intention, he said, was to employ the craft upon America's long rivers, making of them thoroughfares of transportation where roads were nonexistent and where conditions were such as to make the hauling of boats by men or beasts laborious, hazardous and expensive or, in some places, impossible. He evidently had the Mississippi River in mind — not, as Livingston did, the Hudson.

By May of 1803 the hull, fifty-six and a half feet long and ten and a half feet in the beam and lying tied up in the Seine near Perier's plant, was ready to receive Perier's engine, which along with the boat's other machinery was installed sometime that month. The craft, in public view, made quite a spectacle. One Paris journalist described it as having "two large wheels mounted on an axle, like a cart" and behind the wheels rose "a kind of large stove with a pipe." The creation born in Fulton's imaginative mind had at last become a reality, floating as a curiosity on the Seine, nearing completion and the crucial test voyage that would follow.

Then tragedy struck. Fulton was awakened one night to be told that the boat had been damaged and had sunk to the bottom of the river, possibly the work of saboteurs — Seine boatmen bent on destroying what they saw as a threat to their livelihood. Another possible cause, though, was that the heavy machinery had simply broken through a hull inadequate to support it. Fulton quickly roused himself and hurried to the riverfront, enlisting help to recover the machinery from the water. Working through the night and deep into the next day, Fulton and his helpers managed to raise the engine, boiler and condenser and salvage as much else as they could. The boat itself, however, was a ruin, and a new one would have to be built.

By the end of July the new hull was finished, built not with Livingston's money but with funds Fulton raised on his own, probably from Joel Barlow. This time the boat was seventy-four and a half feet long and approximately eight feet in the beam, substantially slimmer than the one that had been destroyed. When all was in readiness, Fulton issued an invitation for a public demonstration of the boat's operation, asking France's prestigious and influential National Institute to send a delegation to witness the demonstration, which was to take place on Tuesday, August 9, 1803, at 6 P.M. The boat was scheduled to make a run in the Seine between the Barriere des Bons Hommes and the Chaillot waterworks, a distance of about a mile.

At the appointed time, a large crowd of curious onlookers gathered on the banks of the river to behold the spectacle to be put on by the strange-looking, fire-breathing, floating contraption that would attempt to navigate the Seine. The event was recorded in the *Journal des Debats*:

At six o'clock in the evening, helped by only three persons, he [Fulton] put the boat in motion with two other boats in tow behind it, and for an hour and a half he afforded the strange spectacle of a boat moved by wheels like a cart, these wheels being provided with paddles or flat plates, and being moved by a fire engine.

As we followed it along the quay, the speed against the current of the Seine seemed to be about that of a rapid pedestrian, that is about 2,400 toises (2.90 miles) an hour; while going down stream it was more rapid. It ascended and descended four times from Les Bons Hommes as far as the Chaillot engine; it was maneuvered with facility, turned to the right and left, came to anchor, started again, and passed by the swimming school.

One of the [towed] boats took to the quay a number of savants and representatives of the Institute, among whom were Citizens Bossut, Carnot, Prony, Perier, Volney, etc. Doubtless they will make a report that will give this discovery all the celebrity it deserves; for this mechanism applied to our rivers, the Seine, the Loire, and the Rhone, would bring the most advantageous consequences to our internal navigation. The tows of barges which now require four months to come from Nantes to Paris would arrive promptly in from ten to fifteen days. The author of this brilliant invention is M. Fulton, an American and a celebrated engineer.[5]

The two boats that Fulton's craft had towed, in which Fulton had provided French government officials and other VIPs a ride to let them participate in the big, historic event, had no doubt slowed his steamboat and had partly accounted for its failure to reach the sixteen-mile-an-hour speed he had predicted. But Fulton was far from being discouraged by the boat's performance. He would simply have to use a more powerful engine next time — and he would give up the idea of towing other craft.

There was a good reason for Fulton and Livingston to move promptly from their prototype, once it had passed its test, as it had, to the actual steam-powered boat that would ply the Hudson on a regular schedule, as Livingston had long envisioned. Livingston had used his and his family's powerful political influence to obtain from the New York State legislature the right to be sole operators of steamboats on the Hudson — a privilege Livingston and Fulton hoped to also gain on other American rivers. The Hudson monopoly had been first granted to Livingston in March 1798, its primary justification being to protect the Livingston boat from potential competition by imitators who would copy the designs of the craft in which Livingston had invested so much of his time and money.

The legislature granted Livingston a monopoly that would, under the terms specified by the legislation, last twenty years, but did so with conditions that the steamboat Livingston proposed must meet. It must: (1) have a capacity of not less than twenty tons; (2) attain a speed of not less than four miles an hour; and (3) be in operation within a year — that is, by March 1799.

The boat that Nicholas Roosevelt had built for Livingston, and Stevens, during its test in October 1798, failed to meet the speed requirement and was eventually abandoned. Livingston managed to have the New York legislature extend its deadline — twice — and the latest deadline was April 1807. He must have a steamboat operating successfully by then in order to keep his monopoly rights.

It was not until April 1804 that Fulton made the move from Paris to pursue the construction of a boat that would meet the New York legislature's conditions. He informed Livingston that he was leaving for England to oversee the manufacture of the Boulton and Watt steam engine that was intended to power a boat that would be built in the United States. He got Livingston to enlist fellow diplomat James Monroe's help in obtaining from the British government the export license needed to ship the proposed engine to America. In so doing, Livingston emphasized to Monroe the importance of the engine to commerce on the Mississippi, over which, thanks in part to their efforts, the United States had so recently acquired complete control through the Louisiana Purchase.

Fulton left for London on April 29, 1804. He did not get to the United States till two and a half years later. Evidently unknown to Livingston, Fulton had been asked by British officials, through an American agent, to come to England to build and stage a demonstration of his submarine, or plunging boat, as it was called, and his mines, which were called torpedoes. Since hostilities between England and France had resumed and a French invasion of England had become a new possibility, the British war office had decided to give Fulton's infernal machines a try in hopes they could be used to combat the menace of the expanding French fleet.

Fulton had leaped at that new chance for fortune and fame. The British government ended up paying him not exactly the fortune he sought but enough to make him richer than he had ever been. His weapons of war, though, failed the test of practicability, and the British government in 1806 at last abandoned its hope of using them to fight France.

Fulton did succeed in obtaining permission for Matthew Boulton (who had taken over the company following Watt's retirement) to build him the needed machinery to his specifications, including the steam engine, the condenser and an air pump, all for the price of 548 pounds, or about $2,740. He also got a London firm to make him a two-ton copper boiler, for 477 pounds, about $2,385. Everything was completed by March 1805, whereupon Fulton was granted a permit to export the parts to America.

With all necessary business taken care of and with no further gains to be realized from his war weapons in England, Fulton finally made preparations to return to the United States. He bought passage aboard a ship and sailed from Falmouth in October 1806. After seven turbulent weeks at sea he arrived safely in New York on December 13, 1806. He had been gone twenty years and was now forty-one years old. With Livingston's deadline of April 1807 just four short months away, Fulton wrote to Livingston shortly after his arrival to tell him that he was ready to move ahead with their steamboat. But then he went to Philadelphia to be with the Barlows, who were living there at the time, and from Philadelphia the three of them went to Washington City, as the nation's capital was then called, and stayed a month, entertaining and being entertained and joining in the celebration that welcomed Meriwether Lewis back from his epic journey across the North American continent.

When he returned to Philadelphia, Fulton dawdled awhile longer, then at last went to New York to get started on the building of the boat. It was the middle of March 1807 when he reported that construction had started at the Charles Browne shipyard at Corlear's Hook, on the East River in lower Manhattan. Not long after that, he took the steam engine and other machinery parts out of storage in the U.S. Customs warehouse, where they had been since November 1806. Near the end of March he reported, "I have now Ship Builders, Blacksmiths and Carpenters occupied at New York in building and executing the machinery of my Steam Boat." He said the boat's construction would take four more months to complete, well past the legislature's deadline. Livingston responded by having the legislature extend the deadline once again, for another two years.

By the middle of July the hull was far enough along for the two paddle wheels to be mounted on its sides and for Fulton to plan a test for the boat at the end of July. The craft was one hundred and forty-six feet long, thirteen feet in the beam, flat-bottomed and straight-sided, but with a curved bow. Two masts, one fore and one aft, would allow the boat to be rigged with square sails that could be used if the engine failed. The cost of the boat's construction had risen sharply, doubling to $10,000 the $5,000 that Fulton had originally estimated he and Livingston each would have to contribute.

The boat's first test was held on Sunday, August 9, 1807, exactly four years since the trial run of Fulton's experimental craft on the Seine. Following the new test, Fulton gave Livingston a brief written report: "I ran about one mile up the East River against a tide of about one mile an hour ... according to my best observations, I went 3 miles an hour.... Much has been proved by this experiment." He told Livingston that he would overhaul the engine and make some adjustments to allow the vessel to make more speed. He predicted that it would achieve the required four miles an hour and said that he was planning to take it on its maiden voyage from New York to Albany on Monday, August 17, 1807.

Even while preparing for the voyage up the Hudson, though, Fulton was thinking ahead. "Whatever may be the fate of steamboats on the Hudson," he told Livingston in the final sentences of his report, "everything is completely proved for the Mississippi."[6] About that same time, he gave an interview to a reporter for the *American Citizen*, who, further revealing Fulton's ambitions, commented in print that Fulton's "Ingenius Steamboat, invented with a view to the navigation of the Mississippi from New Orleans upward ... [would] certainly make an exceedingly valuable acquisition to the commerce of the Western States."[7]

To make sure the boat would be ready for the August 17 run, Fulton put it through another trial on Sunday, August 16. He seemed to prefer Sundays for demonstrations, presumably because Sundays provided bigger crowds and he wanted to show off his creation to as many people as possible. He moved the boat, under steam, from the East River, around the Battery, at the lower tip of Manhattan, then steamed up a short distance on the North River, as the lower Hudson used to be called, and tied the boat up at a dock near Greenwich Village. With him on the short voyage were a number of Livingston's friends and relatives, including United States senator Samuel Latham Mitchell, who when he was a member of the New York legislature had introduced the first bill to grant Livingston a steamboat monopoly on the Hudson.

Monday, August 17, dawned bright and warm, promising a hot summer day, and the tide tables said that high tide, reaching up the river from the lower and upper New York bays, would be at eight o'clock that morning and the tide would flood again just before two that afternoon. With a small crew and about forty passengers, mostly Livingston relatives and friends, aboard with him, and a throng of onlookers lining the docks and riverbank, many of them hooting and jeering the ungainly-looking craft, Fulton shouted orders to his captain, Davis Hunt, and his engineer, George Jackson, and at about one o'clock in the afternoon Fulton's boat, as yet unnamed, cast off its lines and slipped away from the wharf, its grist-mill-looking wheels churning through the water, moving the boat against the current and the ebbing tide, its white-pine fuel sending up columns of smoke and fiery sparks from the craft's tall stacks.

After a short distance, with all seeming to go well, the boat's engine abruptly stopped, setting off a wave of anxiety that swept through the passengers. Fulton described the incident and the reactions to it:

> The moment arrived in which the word was to be given for the boat to move. My friends were in groups on the deck. There was anxiety mixed with fear among them. They were silent, sad, and weary. I read in their looks nothing but

The *North River Steam Boat*, also known as the *Clermont*, the pioneering creation of Robert Fulton and his financial backer, Robert Livingston, steams up the Hudson River from Manhattan and after three days reaches Albany about 5 P.M. on August 19, 1807. "She is unquestionably the most pleasant boat I ever went in," one of its passengers remarked. "In her the mind is free from suspense" (Library of Congress).

disaster, and almost repented of my efforts. The signal was given and the boat moved on a short distance and then stopped and became immovable. To the silence of the preceding moment now succeeded murmurs of discontent, and agitations, and whispers and shrugs. I could hear distinctly repeated — "I told you it was so; it is a foolish scheme; I wish we were well out of it."

I elevated myself upon a platform and addressed the assembly. I stated that I knew not what was the matter, but if they would be quiet and indulge me for half an hour, I would either go on or abandon the voyage for that time. This short respite was conceded without objection. I went below and examined the machinery, and discovered that the cause was a slight maladjustment of some of the work. In a short time it was obviated. The boat was again put in motion. She continued to move on. All were still incredulous. None seemed willing to trust the evidence of their own senses. We left the fair city of New York.[8]

From that time on, the voyage — and the steamboat — proceeded flawlessly. Onward the craft chugged and splashed, and as night came on, the women retired to a cabin near the stern. On cots provided them they tried to sleep despite the constant pounding of the engine. The men, most of them, stayed up, keeping themselves occupied by singing, their voices carrying across the water in the night air. They passed through Haverstraw Bay, churned past the

Highlands and on beyond West Point. At daybreak the light revealed the many curious onlookers along the riverside. About the middle of the morning of Tuesday, August 18, the boat came round the river's bend at Kingston, and the passengers could then make out the Catskill Mountains in the hazy distance.

At last, at one o'clock that afternoon, they reached the dock at Livingston's Clermont estate, the one scheduled stop on the trip to Albany. Its safe arrival signaled a huge success for the boat. It had made the one-hundred-and-ten-mile voyage to Clermont in twenty-four hours, achieving an average speed of slightly better than four and a half miles an hour. Waiting at the dock was a jubilant Robert Livingston, happy to see and eager to welcome the pioneer voyagers and the novel craft that had brought them to him.

At a little past nine o'clock the next morning, Wednesday, August 19, having left most of his passengers at Clermont, Fulton steamed off again. This time, Livingston was aboard. The boat, which later would be named simply the *North River Steam Boat*, but would become popularly, although mistakenly, known as the *Clermont*, arrived at Albany about five P.M. after an uneventful forty-mile trip. The governor of New York and a host of astonished citizens were at the waterfront to offer an enthusiastic welcome to the participants of the history-making voyage.

One of the passengers, an Anglican clergyman, reporting on the experience, summed up the remarkable success of Fulton's boat: "She is unquestionably the most pleasant boat I ever went in. In her the mind is free from suspense. Perpetual motion authorises you to calculate on a certain time to land; her works move with all the facility of a clock; and the noise when on board is not greater than that of a vessel sailing with a good breeze."[9]

The next morning, Thursday, August 20, 1807, Fulton and his boat were ready to leave for the return trip to New York. This leg of the *North River Steam Boat*'s round trip, however, would not be a free ride. The boat overnight had become a commercial vessel. Fulton had a sign made and had it hung on the side of the boat. The sign announced that passage to New York was available to the public. The fare: seven dollars, meals included. Two courageous travelers bought tickets to become the *North River*'s first paying passengers.

America's steamboat era had begun.

• 5 •

A Different Kind of Boat

When Robert Fulton at last persuaded Robert Livingston to lift his eyes and see the great opportunity that lay beyond the Northeast, Nicholas J. Roosevelt was the man the partners picked to build a steamboat that would voyage down the Mississippi River and become the first one to do so. Fulton and Livingston's objective was to acquire a steamboat monopoly on the lower Mississippi as they had on the Hudson and thereby establish control of commercial shipping on two of the country's most important waterways. But before asking the Louisiana territorial government to grant them a monopoly and before building the steamboat, Fulton and Livingston wanted to make sure the boat they envisioned could indeed make the voyage across the length of the Ohio and down the Mississippi all the way to New Orleans. To assure themselves, they proposed to send Roosevelt to Pittsburgh to have a boat built and make a test run that, if successful, would prepare the way for the coming of the steamboat. The test run would be made in a flatboat, propelled mainly by the rivers' currents.

Roosevelt (whose brother became the grandfather of Theodore Roosevelt, the twenty-sixth president of the United States) was, like Livingston, from an old, respected New York family. His father, Isaac, had been a member of the New York legislature and was for many years president of the Bank of New York. Roosevelt's foundry and boat works, however, where he had built the experimental steamboat for Livingston and John Stevens, had fallen on hard times, and Roosevelt had become open to new propositions.

In the spring of 1809 he was forty-one years old and recently married to eighteen-year-old Lydia Latrobe, with whom he had fallen in love when she was a precocious fourteen-year-old. High-spirited and strong-minded, she was the daughter of architect, engineer and inventor Benjamin Latrobe, with whom Roosevelt had participated in several business ventures and who had become Roosevelt's close friend. Agreeing to help Fulton and Livingston pursue their plan for a Mississippi River steamboat, Roosevelt would make the rugged journey over the Alleghenies to Pittsburgh to build a boat aboard which he would drift through the heart of America, down to New Orleans. With him would go his courageous, venturesome young wife, who would not let him go without her. The trip would be for them, still newlyweds, like a honeymoon voyage.

They set out for Pittsburgh in the spring of 1809 and upon their arrival there immediately began to design their flatboat and arrange to have it built. Pittsburgh was then a town of about four thousand residents and a place where boat-building was a thriving industry, many freight-carrying flatboats originating there and there beginning their one-way voyage down the river.

The boat that Nicholas — and Lydia, the architect's daughter — designed was essentially a houseboat with two cabins. The aft cabin — which Lydia called "a huge box" — contained a bedroom, a dining room and a pantry for the couple. The forward cabin, the larger of the two, housed the five-man crew — the vessel's pilot, a cook and three deckhands, one of whom would man the tiller and two who would man the sweeps, the long-handled oars that provided additional propulsion when needed. The forward cabin also included a stone or brick fireplace where the cooking would be done. The top of the boat formed a flat, upper deck that was covered by an awning and had seats on it.

The boat carried or towed a large rowboat from which soundings could be taken to determine the depth of the water and the speed of the current in advance of the boat's reaching shoals, eddies, white water or dangerous-looking obstructions. Those measurements and the locations at which they were taken would be meticulously recorded by Roosevelt in a notebook he would keep throughout the voyage and in which he would also draw maps so that they and the notes could be referred to when the proposed steamboat, with a deeper draft, would later venture into the same waters.

In June 1809 the Roosevelts' flatboat was ready for them, and they cast off from the dock at Pittsburgh to start their two-thousand-mile journey of exploration, gliding down to the confluence of the Monongahela and Allegheny rivers and entering the Ohio at its beginning. They traveled by day and tied up to the riverbank each night as darkness came on. Nicholas soon found that his days would not be leisurely spent with Lydia beneath the awning of the upper deck. Instead, he was usually out in the rowboat taking soundings, making observations and taking notes.

The first major port of call was Cincinnati, about four hundred and fifty river miles from Pittsburgh, a town of about twenty-five hundred residents. Nicholas and Lydia were greeted and entertained by some of the town's officials and leading citizens and after a short stay, they resumed their voyage. The next significant stop was at Louisville, about a hundred and thirty miles farther downriver, a community of about twelve hundred persons. Again they were warmly received. They were also warned of a serious danger that lay ahead of them — the Falls of the Ohio, a stretch of the river that dropped twenty-six feet over a three-mile distance, creating rapids that coursed over and through menacing rocks and stone ledges against which their boat could be dashed to splinters. The warning gave Nicholas pause. He took three weeks to study the falls, measuring distances and the depth and speed of the water.

The Roosevelts' stay in Louisville lasted long enough for them to make friends in the community, friends they would see again. As they had at Cincinnati, they told people what they were doing — scouting out the Ohio and Mississippi in preparation for running a steamboat on the rivers. And as at Cincinnati, they were met by much skepticism, the popular belief being that although a boat driven solely by a steam engine might make it *down* the rivers, it could never travel *up* them, against the currents. When at last Nicholas was satisfied that his study was complete, he hired a special pilot, as most boatmen did, to steer him through the falls and he and Lydia and their crew, now joined by the local pilot, boarded the flatboat and shoved off to brave the perilous falls.

They were as fearsome as had been expected, but the pilot succeeded in getting the boat through them, and once they were passed, the river resumed its tranquility, and the party of exploration continued on its way, passing through a world of woods that bordered the river on both banks. Only occasionally did they see human life, isolated settlers who came out to the water's edge to hail them, or the crewmen of other boats making their way downstream. Some days the Roosevelt boat stopped to let its crew fish or hunt to replenish the food supply. One day Nicholas discovered two beds of coal on the riverbank, about one hundred and twenty miles below the falls, and he made a note of their location so that a quantity of coal could be dug out later and used as fuel for the planned steamboat. Most days, though, the boat pushed slowly on, and Nicholas continued to measure and sound and make entries in his notebook.

When they had come a distance that Nicholas estimated to be a thousand miles from Pittsburgh (it's actually slightly less), the boat reached the mile-wide Mississippi, and the crew deftly steered the flatboat out into the muddy mainstream of the mighty river, the blue water of the Ohio turning gray as it was absorbed into the larger flow. With the entry into the Mississippi came new dangers, including floating and imbedded objects of many descriptions, particularly fallen trees that created snags in the stream, any of which might smash or rip open the boat's hull, and, not the least threat, hostile Indians who could reach the boat in their canoes, dangers that their friends in Louisville had warned them about. One night the boat was indeed boarded by Indians. Lydia wrote about the incident: "Mr. Roosevelt was aroused in the night by seeing two Indians in our sleeping room, calling for whiskey, when Mr. Roosevelt had to get up and give it to them before he could induce them to leave the boat."[1]

Stopping only briefly at New Madrid, a town on the west bank of the river in the Missouri territory, the Roosevelts and their crew continued languidly down the broad Mississippi, slowly slipping southward, past great wooded stretches and occasional cultivated fields, hearing the sounds of birds and wild animals breaking the encompassing silence of the river. Deeper into the southland they drifted, where the land became more settled looking, with fields stretching away from the river. Then at last they came upon Natchez, the great cotton gathering center, its riverfront crowded with motley boats and rough shacks, its handsome commercial and residential areas standing serenely aloof on the hill that rises from the water's edge. At Natchez they were heartily greeted by the town's luminaries and there they received an offer for their flatboat, one that was, Nicholas felt, too good to refuse, unlikely as it was to be matched in New Orleans. Since they were now within days of their destination, Nicholas decided to sell the boat that had been their comfortable home for so many weeks and make the rest of the voyage in the large, open rowboat from which he had been making his soundings and observations.

The honeymoon was now over. "Our pilot," Lydia reported, "who had lived all his life as a boatman on these waters, assured us that there would be no difficulty in finding lodgings for the few nights we should be out. But it appeared the inhabitants on the river had been so often imposed on by travelers whom they had received into their houses, that they refused all applications."[2] After long, tiresome days in the rowboat, unprotected against the sun and weather, the couple did the best they could to rest at night. They spread a buffalo robe across a large trunk at the stern of the boat to make a bed and spent four nights trying to sleep that way in the boat, which was partly drawn out of the river. All the while, Lydia recalled, they were "hearing the alligators scratch on the sides [of the boat], taking it for a log; [and] when a knock with a cane would alarm them..., they would splash down into the water."[3]

Three nights they spread their buffalo robe on the sandy river bank and tried to rest there, but "feeling every moment," Lydia said, "that something terrible might happen before morning."[4] Two other nights they managed to find shelter inside buildings, one the cabin of "an old French couple who allowed us to spread our buffalo robes on the floor before a fine, large fire, where we felt safe," Lydia wrote, "though disturbed once or twice during the night by the people coming into the room we occupied, and kneeling before a crucifix which stood upon a shelf."[5] The other building in which they found a night's lodging was a tavern in Baton Rouge. The room, Lydia reported, "was a forlorn little place opening out of the bar-room, which was filled with tipsy men looking like cut-throats. The room had one window opening into a stable-yard, but which had neither shutters nor fastenings. Its furniture was a single chair and dirty bed. We threw our cloaks on the bed and laid down to rest, but not to sleep, for the fighting and the noise in the bar-room prevented that. We rose at the dawn of day, and reached the boat, feeling thankful we had not been murdered in the night. It is many, many years ago; but I can still recall that night of fright."[6]

The exploration party finally reached New Orleans on December 1, 1809, but had little time to linger and recover. Instead, they sought out the first available ship leaving for the East Coast and quickly sailed away in it. That voyage, like the nine days the Roosevelts had spent in the open rowboat, was not much of a honeymoon either. "We had a terrible voyage of a month, with a sick captain," Lydia wrote. "The yellow fever was on board. A passenger ... died with it." The Roosevelts left the ship off the coast of Virginia and were taken by a pilot boat to Old Point Comfort and from there they took a stagecoach to New York, arriving in mid–January 1810. They had been gone nine months.

Nicholas promptly reported to Fulton and Livingston what he had learned. With that good news in hand, Fulton and Livingston entered into a contract with Nicholas. Some sources say the arrangement was a partnership that provided for Fulton and Livingston to supply the capital and Nicholas the expertise and time. Under the agreement, Nicholas would go back to Pittsburgh and there oversee the building of a steamboat according to Fulton's specifications. Also under the terms of the agreement, Nicholas would take the steamboat down the Ohio and the Mississippi to New Orleans, retracing the journey he and Lydia had made in the flatboat. In honor of its city of destination, the new steamboat would be named the *New Orleans*.

Fulton's plans called for a side-wheeler one hundred and sixteen feet long and twenty feet in the beam, drawing about seven feet of water (a specification which Nicholas was said to alter in order to reduce the vessel's draft). Its hull would be rounded, like that of a seagoing vessel. The engine was to have a thirty-four-inch cylinder with an appropriately sized boiler mounted in the vessel's hold. The *New Orleans* would have two cabins, one aft for women passengers and a larger one forward for men. The women's cabin would contain four berths and would be comfortably furnished. According to one account, the vessel would have portholes and a bowsprit and would be painted light blue. It would have two masts and carry sails for use if needed. Timbers for the boat's ribs, beams and knees would come from forests near Pittsburgh, the felled trees to be dragged into the Monongahela River and rafted downstream to the site where the boat would be constructed.

It would be built on the bank of the Monongahela, near Boyd's Hill, at a location on which was later erected the depot of the Pittsburgh and Connellsville Railroad, close to Beelen's foundry. Mechanics to build and install the engine and other mechanical equipment were brought from New York, but local boatwrights, under Nicholas's oversight, would assemble the craft. The cost of the vessel would come to about $38,000, which Livingston thought excessive.

On September 27, 1811, the *New Orleans* was at last finished and ready to be launched on its history-making voyage. On board were Nicholas and Lydia, a captain whom the records leave unnamed, an engineer named Baker, a pilot named Andrew Jack, six deckhands, two maids, a male waiter, a cook and an enormous Newfoundland dog named Tiger. Friends of the Roosevelts implored Lydia, who was eight months pregnant, not to go, but she was determined to make the trip. Pittsburgh's residents turned out in huge numbers to see them off and to see if the steamboat would actually work. They thronged along the banks of the Monongahela, waving handkerchiefs, tossing their hats into the air and shouting as the *New Orleans* shoved off, smoke rising like tall clouds from its two black smokestacks. It glided down to the confluence and at last disappeared from the crowd's view as it passed behind the headlands on the west bank of the Ohio.

Coincidental to the momentous voyage of the first steamboat on America's western waters was the strange appearance of a comet that was visible to the naked eye for most of the year (much like the Hale-Bopp comet of 1997), setting off waves of consternation in the hearts of many who saw and feared it. If the comet, known simply as C/1811 F1, was the portent of a coming calamity, as some believed it was, Nicholas and Lydia and the crew of the *New Orleans* would soon enough learn what the calamity was.

As the boat's only passengers, Nicholas and Lydia had the cabins to themselves, but too excited to sleep, they spent most of the first night on deck watching the forested riverbanks, shadowy in the moonlight, slip by them while the boat proceeded downstream at a speed of eight to ten miles an hour.

During the second night out of Pittsburgh the *New Orleans* reached Cincinnati and dropped its anchor in the stream. A crowd of Cincinnati's citizens was up waiting for the Roosevelts, many of them having become acquaintances when Nicholas and Lydia stopped there earlier in their flatboat. Some came out in rowboats to greet them, telling Nicholas and Lydia, "Well, you are as good as your word. You have visited us with a steamboat." But the doubters were steadfast in their belief that the steamboat could not successfully buck the river's current and move upstream. "We see you for the last time," some were reported to say. "Your boat may go down the river, but as to coming up, the idea is an absurd one."[7]

Keelboat crewmen in Cincinnati were among the most outspoken unbelievers, hurling gibes at the crew of the *New Orleans*. A number of flatboatmen who had watched the *New Orleans* steam past them just upstream of Cincinnati showed a bit more respect for the vessel, proposing that the steamer give them a tow if it passed them again. But they, too, refused to believe the *New Orleans* could make it upstream.

After taking on a supply of wood fuel, the boat steamed off again, headed now for Louisville, which it reached about midnight on October 1, four days out of Pittsburgh. It anchored opposite the town, under a brilliant moon. When the boat's engineer opened the valve to let off steam and stop the engine, the escaping steam made such a loud and strange noise that it awakened the townspeople, who despite the late hour, came swarming to the riverfront to behold the fire-breathing, floating monster, clearly visible in the bright moonlight. One of the *New Orleans* crewmen later wrote a letter in which he claimed that the people of Louisville had been convinced that the 1811 comet had fallen into the Ohio and was the cause of all the commotion.

Several days after the Roosevelts arrived in Louisville, a public dinner was given for them, and Nicholas was hailed and saluted with toasts in celebration of his accomplishment in building the steamboat and bringing it to Louisville for all to see. It was a heartwarming occasion. However, before the evening passed, comments were inevitably made about the chances of Nicholas's boat being able to return upstream. Some expressed their regret that this was not only the first but also the last time a steamboat would be seen above the Falls of the Ohio.

Nicholas took it all in good humor and then graciously invited all his hosts to be his guests at a dinner one evening aboard the *New Orleans*. They accepted and gathered for the shipboard *soiree* in the gentlemen's cabin. Midway through the banquet, the conviviality was alarmingly interrupted by rumblings from below, and the guests felt the vessel begin to move. Suspecting the boat had somehow lost its anchor and was now adrift, headed for the Falls of the Ohio, the diners rushed to the boat's upper deck to confirm their fears. What they saw, however, was that they were headed upstream, the powerful paddle wheels of the *New Orleans* churning steadily against the Ohio's current. Within minutes the riverfront of Louisville was downstream of them, fading into the distance. After continuing a few miles up the river, the boat turned around and returned to its anchorage opposite the town. Nicholas had staged a convincing demonstration of what his steamboat could do.

When the Roosevelts were ready to leave Louisville, Nicholas learned that the river at the Falls of the Ohio was not deep enough for the *New Orleans* to get through safely. And so while waiting for the river to rise to a sufficient depth, he took the vessel back to Cincinnati and made more believers of the steamboat's prowess. He then returned to Louisville, where, while the wait continued, Lydia gave birth. She had expected the baby to be born aboard the *New Orleans*, but during the wait she had accepted the hospitality of Louisville friends, and the baby was born in the friends' home.

It was not until the last week of November that the river rose enough for Nicholas to risk the passage of the *New Orleans* through the rapids of the Falls. To help get the boat through, Nicholas took on two special pilots, who stationed themselves at the bow of the boat, studying the frothing river before them. Lydia watched from the stern. Hugging the Indiana side of the river, the *New Orleans* sped through the menacing rapids.

The danger and fright of the Falls of the Ohio were now behind the voyagers. Peril even more terrifying, however, lay ahead of them.

First there was an attempt by Indians to catch the boat. A large canoe, fully manned, suddenly darted out from the woods beside the river and quickly came up behind the *New Orleans*, the Indians paddling furiously to overtake it in a race that the steamboat won when the Indians, their arms eventually tiring, issued a barrage of wild shouts and quit the chase.

Then there was the fire aboard the boat. One of the servants had stacked wet wood near the stove in the forward cabin in an effort to have it dry quickly. The wood was placed so close to the stove that it overheated and caught fire, spreading flames to the woodwork in the cabin, which were soon extinguished by the crewmen before major damage could be done to the vessel.

When the *New Orleans* reached the spot on the Indiana side of the river where Nicholas had noticed the coal vein, which he had in the meantime purchased from the government, the vessel pulled over to the shore to dig out the coal and take it aboard. They found that a large quantity of the coal had already been dug out of the vein and had been piled up near the river bank. Since the coal actually now belonged to Nicholas, he had his crewmen load it onto the boat. While the loading was being done, several people, described by one nineteenth-century account[8] as "squatters of the neighborhood," apparently frightened, came up to Nicholas and his workers and asked if they had heard strange noises on the river and in the woods during the day before. The squatters reported that they had repeatedly felt the earth tremble beneath them and had seen the bank of the river shake.

Those aboard the *New Orleans* evidently *had* felt a tremor as the boat came down the river, but it was not until the day after they had stopped for the coal that, with repeated shocks, they realized something horrific was happening. The weather turned oppressively hot, the air misty, the mid-day sun a copper-colored ball that shone but dimly on the surface of the river, as if at twilight. Sitting on the upper deck as they resumed the voyage, the Roosevelts and their fellow travelers could hear an occasional rushing sound and then a violent splash and they saw great chunks of the shore rip away and fall into the river. And yet, except for the occasional splashing of earth into the water, an eerie silence enveloped the river and nearby woods. Everyone aboard the boat seemed stricken with fright and wonder.

The next day was the same. The pilot, Andrew Jack, confessed he had no idea where they were on the river. Alarmed and confused, he reported that the Ohio's channel was completely changed. Landmark trees and bluffs that had been his guides along the course of the river were now gone, vanished into the Ohio's muddied, yellow waters. Where there had been deep water there were now uprooted trees lying thick in the stream. Islands in the river that had served as guides had disappeared or had been turned into unrecognizable shapes. Menacing reefs and bars had suddenly appeared where the current had once flowed unobstructed. The *New Orleans* was steaming in uncharted waters, through a mysteriously altered landscape.

There was, however, no alternative but to press on. As evening approached on that second day from the coal vein, the voyagers sought a place where they could find shelter from the river's current and tie up safely for the night, as they had been doing. Now they could find no such place. They could see flatboats and rafts that had similarly sought a place to tie up and were partly covered by falling earth where bluffs had caved in and slid on top of the vessels, now abandoned by their crews. The pilot, Jack, decided to continue on to a large island with which he was familiar, one that stood in mid-channel and that offered places to tie up. He could not find it, though, it evidently having disappeared. Tensely they proceeded onward, the passing hours slowly taking them into darkness. Finally they came upon a small island and moored the *New Orleans* at the foot of it. There they passed a nervous night, sleeping little, listening to the river rush past them and the sounds of earth and trees falling into the stream, and watchful of the furniture being sent skidding across the deck of the cabin as the boat was violently struck and jarred by debris floating in the river.

When the reassuring light of morning came at last, the voyagers could see that they were very near the mouth of the Ohio and that the Mississippi lay not far in the distance, even though the banks of the river and the river channels were now completely changed. Soon the *New Orleans* made a wide southward turn and entered the Mississippi's flow. Not long after that, the voyagers reached New Madrid, which in 1811 was the most important town on the Mississippi River between St. Louis and Natchez.

At New Madrid the Roosevelts and their fellow travelers aboard the *New Orleans* got some idea of the nature of the calamity into which they had haplessly sailed. New Madrid was very near the epicenter of an awesomely powerful and far-reaching earthquake, the biggest ever to hit North America and which has become known as the New Madrid Earthquake. It was felt as far away as Washington, D.C.; Raleigh, North Carolina; and Savannah, Georgia, as well as in many places in between.

The earthquake caused a large section of forested land in the northwestern corner of Tennessee to sink below the level of the river, and the Mississippi rushed backwards for hours to fill the enormous hole. Once the giant depression had been filled with water, the river resumed its southward flow. The depression, still filled with water, remains a natural feature of the land in northwest Tennessee, about twenty miles from present-day New Madrid, and is called Reelfoot Lake.

The earthquake had left New Madrid in ruins. Many of the houses and other buildings had been swallowed by the earth as fissures opened up or had been swept away after the land beneath them had fallen into the river. The town's cemetery had also fallen into the river. At one site, according to one account, the earth's upheaval had uncovered the fossilized bones of a mastodon. Some of the town's residents had felled trees perpendicular to the cracks in the ground and were clinging to them in hopes the fallen trees would bridge the fissures and prevent them for falling into the chasms. Some residents had fled the town, seeking higher ground, but many were still there as the *New Orleans* approached, giving some of them a new fright. Many others among the quake's survivors hailed the boat and begged to be taken aboard to escape the town. Lacking provisions sufficient for a large increase in passengers, however, Nicholas had to refuse them.

The *New Orleans* proceeded down the swollen Mississippi, fallen trees floating all around the boat, the pilot trusting more to luck than to an ability to read the altered landscape and current patterns, choosing the flow's strongest current in his efforts to find the changed channel of the river.

At the end of the first week of January 1812, the perils of the earthquake's effects behind them, the intrepid voyagers reached Natchez. Then came a happy occasion. During the voyage from Pittsburgh the boat's captain and one of Lydia's maids had fallen in love, and she had accepted his proposal of marriage. When the vessel docked at Natchez, Nicholas hunted up a clergyman, and a hastily arranged wedding ceremony was held for the couple. Before it left Natchez, the *New Orleans* took on a shipment of cotton — although the shipper's friends warned him against putting it on the steamboat — for delivery in New Orleans. That cotton thus became the first steamboat freight on the Mississippi.

On January 12, 1812, welcomed by thousands of cheering onlookers who crowded the levee to get a look, the *New Orleans* at last arrived at the city for which it was named, its history-making voyage ended, more than three months and more than two thousand miles after it had begun.

Nicholas was left in charge of the *New Orleans* and operated it on its scheduled runs between New Orleans and Natchez. It made the round trip regularly in seventeen days, carrying freight and passengers, and in its first year of operation it earned a profit of $20,000. One source[9] quoting from a publication of the early nineteenth century, gives details — some presenting a reporter's odd arithmetic — of the boat's operations:

> Her accommodations are good, and her passengers numerous, generally not less than from ten to twenty from Natchez at $18.00 each, and when she starts from New Orleans, generally from thirty to fifty and sometimes as many as eighty, at $25.00 each to Natchez....
> She performs thirteen trips in the year, which at $2,400 amounts to $31,200. Her expenses are, 12 hands at $20 per month, $4,320 [sic]; captain, $1,000; seventy cords of wood each trip, at $1.75, which amounts to $1,586 [sic], in all $6,906. It is presumed that the boat's extra trip for pleasure or otherwise, out of her usual trade, have paid for all her repairs, and with the bar-room, for the boat's provisions....
> She goes up in seven or eight days, and descends in two or three, stopping several times for freight and passengers. She stays at the extreme of her journey, Natchez and New Orleans, about four or five days to discharge or to take in loading.

Before long, Nicholas had a falling out with Fulton and Livingston, apparently over his failure to give the partners regular reports of the boat's operations, and Fulton sent his wife's brother, John Livingston, to New Orleans to take over the boat's records and the boat itself. At first Nicholas refused to turn over the boat, but when John Livingston threatened a lawsuit, Nicholas gave in, and Livingston assumed possession of the *New Orleans* and took charge of its operation.

Nicholas and Lydia eventually reached a financial settlement with the Fulton-Livingston partnership, compensating Nicholas for his contributions to the partners' steamboat success, and the couple later moved to the quiet little town of Skaneateles in the picturesque Finger Lakes area of upstate New

York. There they lived until Nicholas's death on July 30, 1854, at age eighty-six. Lydia died in 1871, at age eighty.

The *New Orleans*, the steamboat that had made them famous and had conquered the mighty Mississippi, lasted not nearly so long. It had a hole punched in its hull when it became impaled on a stump and it sank near Baton Rouge on July 14, 1814.

· 6 ·

Captain Shreve's Design

By the time the *New Orleans* wrecked and sank, the Fulton-Livingston partnership had put another steamboat on the Mississippi — the *Vesuvius*, one hundred and sixty feet long, thirty feet in the beam and drawing six feet of water. It was built in Pittsburgh under the supervision of John Livingston and while steaming down the Ohio on the way to New Orleans on its maiden voyage it achieved an average speed of ten and a half miles an hour. It arrived in New Orleans in May 1814. Fulton had intended it to run between New Orleans and Louisville, but when its hull, like that of the *New Orleans*, proved too deep for the vessel to regularly navigate the shallow waters of the Mississippi above Natchez, Fulton had to change the plan.

The proof of its inability to navigate shallow western waters came dramatically when *Vesuvius* ran aground off the Tennessee shore on its first trip from New Orleans to Louisville and lay like a beached whale for five months before it could be refloated. After that, its service was limited to the New Orleans-to-Natchez run, through deeper water. Despite all his engineering powers, Fulton was having difficulty understanding that boats designed like deep-hulled seagoing vessels, although satisfactory for rivers in the East, would not work well on the Mississippi and other western rivers. They simply drew too much water.

Fulton and Robert Livingston, even before the *New Orleans* made its historic maiden voyage, had gained from Louisiana the monopoly they wanted. They were granted exclusive rights to steamboat navigation on the Mississippi for eighteen years, and although their privilege did not extend beyond the limits of the present-day state of Louisiana, it did include the all-important stretch of river that gave access to the port of New Orleans, a situation that was only bound to be challenged. Daniel French, a Pittsburgh inventor and entrepreneur, was the first to do so.

French headed up a group of investors that built two small steamboats, the *Comet* and the *Despatch*. In an attempt to cope with shallow rivers, French built his boats small enough and light enough to avoid drawing a lot of water. The *Comet* was only fifty-two feet long and eight feet in the beam. He also put a more powerful engine in them. Unlike Fulton's vessels, which used low-pressure steam engines, the *Comet* was driven by a high-pressure engine, which French had designed. In 1813 he launched the *Comet* into the Ohio and sent it down to New Orleans. Sources vary on whether the Fulton-Livingston partnership chased the *Comet* away with a threat of seizure or simply ignored it. In any case, the *Comet* withdrew to Natchez. It turned out that despite its small size it could not dependably navigate above Natchez, and French gave up on it, removing its engine and selling it to run a cotton gin.

French then built a bigger boat, the *Enterprise*, eighty feet long and twenty-nine feet in the beam,[1] and in late 1814 he hired twenty-nine-year-old Henry Miller Shreve to captain it and take it to New Orleans, defying the Fulton-Livingston monopoly.

Shreve had already shown himself to be a remarkable young man and in time to come, he would prove even more remarkable. Born October 21, 1785, he was one of four sons of Israel Shreve and the fifth child of Israel's second wife, Mary Cokely Shreve. At the time of Henry's birth, Israel and Mary and their family were taking refuge at Israel's brother's home on Rancocas Creek in New Jersey following a fire that had destroyed their house. In 1788 the family moved to western Pennsylvania to live on land bought from George Washington, with whom Israel had served as a colonel in the Revolutionary War.

Near the family's new homesite, southeast of Pittsburgh, flowed the Youghiogheny River, a tributary of the Monongahela, and young Henry developed a fascination for it and for the distant places to which it might take him. When his father died in 1802, Henry went to work as a flatboat crewman and learned how to handle riverboats and how to make money running one. In the summer of 1807, when he was twenty-one years old, nearly six feet tall, slender and sinewy, he began building a boat of his own, a keelboat with which he planned to go into business carrying trade goods on the Ohio and Mississippi. In October 1807 the boat, built at Brownsville, Pennsylvania, on the Monongahela, was finished, and Henry recruited a ten-man crew from the Brownsville riverfront to man the boat on a voyage down the Ohio and up the Mississippi to St. Louis. After stopping at Pittsburgh to buy an assortment of goods to sell in St. Louis, he began the descent of the Ohio. At the Ohio's juncture with the Mississippi, his crewmen took to their sweeps and laboriously rowed upstream, reaching St. Louis six weeks after having left Pittsburgh. Shreve sold the goods he had brought to St. Louis, then bought a load of furs and headed his keelboat down the Mississippi, up the Ohio and back to Pittsburgh, where he loaded the furs onto wagons and shipped them to buyers in Philadelphia. That done, he took the boat back to St. Louis to repeat the process.

After three years of trading in furs, Shreve in 1810 tried a new venture, taking his boat up the Mississippi past St. Louis and into Indian country in Illinois, where white men other than British traders from Canada rarely traveled. With a shipload of goods he believed would appeal to the Indians — farming implements, metal pots and a variety of hardware — he and his crew slowly made their way up to the mouth of the Galena River, in the extreme northwestern corner of present-day Illinois, and entered the Galena. Fourteen days later, working the boat upstream, they arrived at a Sac and Fox Indian village that was the base of a Sac and Fox lead mine. Winning over the wary Indians, Shreve traded his goods for their smelted lead and loaded his boat with it. He built another boat there on the site, bought a third boat from another trader and loaded both of them with lead also. On July 1, 1810, he told his new Indian friends goodbye and shoved off for the Mississippi with his three boats and their cargo.

Riding the current, Shreve and his little flotilla swiftly descended fifteen hundred miles of the Mississippi to New Orleans. There he found a sailing ship that would take him and his cargo to Philadelphia, where he arrived that fall and sold the lead for an eleven-thousand-dollar profit. He then returned to Brownsville to have a new boat built and to win the hand of the girl with whom he had fallen in love, red-haired, nineteen-year-old Mary Blair. He married her in February 1811. Within a few months after the wedding he was back on the river, taking his new vessel, a capacious barge, and its freight to New Orleans.

It was apparently on that voyage that Shreve became keenly interested in steamboats. On his return trip from New Orleans he arrived in Louisville in time to see — and examine — the *New Orleans,* moored there while Nicholas Roosevelt waited for the Ohio to rise enough to pass the falls. Shreve was captured by the steamboat. Its possibilities were immediately obvious. No more would a riverboat have to be rowed, poled, warped or towed by crewmen trudging doggedly along the banks of the river, tortoise-slow, tedious and back-breaking work, but until the coming of the *New Orleans,* the only ways to move boats of size against the big river's unremitting current.

Shreve made several more round trips between Brownsville and New Orleans in his own boat before he signed on to captain Daniel French's *Enterprise,* a job that would give Shreve valuable experience running a steamboat, which was exactly what he wanted.

By that time, late 1814, the War of 1812 had been going on for more than two years, and the Mississippi River had become an important part of England's strategy to contain — and perhaps regain — the United States. English warships were blockading America's Atlantic and gulf coasts, and the Mississippi and its tributaries had become more vital than ever as a communication line and as a conduit for moving the nation's goods. England's plan in 1814 was to launch an invasion from the Gulf of Mexico and capture New Orleans, thereby gaining a stranglehold on the Mississippi. The British could then control the big river from Canada and the upper Midwest all the way to the gulf. The United States would be hemmed in, checked by enemy forces on all four sides — Canada to the north, the gulf to the south, the Atlantic to the east and the Mississippi to the west — and it would be forced to submit to whatever terms England might contrive.

On December 1, 1814, Major General Andrew Jackson arrived in New Orleans to take command of the city's defense and soon set about assembling a rag-tag army made up of U.S. regular army troops, U.S. Navy sailors and Marines, Tennessee and Louisiana militiamen, Kentucky frontiersmen, free blacks, Choctaw Indians, an assortment of volunteers and Jean Lafitte's buccaneers, the pirates of Barataria. To stop the invaders before they reached New Orleans, Jackson began erecting a first line of defense that would stand practically in the face of the enemy. He would position his infantry, including the long-rifle marksmen from Kentucky, and his artillery, including guns manned by Lafitte's expert cannoneers, behind a protective wall of cotton bales and mud, stretching from the levee, across the cane field and into the swamp on the east side of the field.

On December 13 the fleet carrying the British invasion force arrived at Lake Borgne, a bay of the gulf, east of the mouth of the Mississippi. The British troops were disembarked into small boats and were rowed across Lake Borgne and up Bayou Bienvenue to reach the rear of the sugar-cane plantations that fronted on the Mississippi some ten miles below New Orleans. From there they planned to march on New Orleans.

Shreve had left Pittsburgh aboard the *Enterprise* on December 1, 1814, and he arrived at New Orleans on December 14. His cargo this time was artillery and ammunition for General Jackson's army, still gathering in and around New Orleans. One of the first to learn of the *Enterprise*'s arrival was Robert Livingston's younger brother Edward, who had moved to New Orleans from New York some years earlier. Robert Livingston had suffered a fatal stroke on February 25, 1813, and his death had thrown his steamboat interests into chaotic disarray, scattered among heirs and others. Out of the disarray eventually had come a new corporation that succeeded to the partnership's

Mississippi River steamboat monopoly and in which Edward Livingston was a major shareholder. A skillful lawyer, he immediately moved to have the *Enterprise* seized on the grounds that it was in violation of the corporation's steamboat monopoly.

He was not quick enough. Before Livingston could obtain a court order for the boat's seizure, General Jackson commandeered the *Enterprise*. He ordered Shreve, once its cargo had been discharged, to take the *Enterprise* up the river and find three keelboats that were supposed to be bringing a shipment of desperately needed small arms to supply Jackson's army.

Preparatory to hunting down the three keelboats, Shreve was making repairs on the *Enterprise* when he was suddenly confronted by a number of marshals who boarded the vessel as it lay beside the wharf. They read a court order to him, told him they were seizing the boat and ordered him off. Shreve replied that he was unable to surrender the boat to them because it had been commandeered by General Jackson and he, Shreve, was under orders from the general to take the boat on a vital military mission upriver. If they wanted the *Enterprise*, they would have to see General Jackson. Defeated, the marshals withdrew. Livingston had been thwarted, for the time being.

Shreve then took the *Enterprise* up the river, searching for the errant keelboats, and found them north of Natchez. He brought their captains aboard the *Enterprise*, tied the keelboats to the *Enterprise* with tow lines and hauled them into New Orleans, arriving a week after he had left.

The all-out British attack came early in the morning on January 8, 1815, setting off a ferocious response from the American line. The defenders' artillery and rifle fire cut like a scythe through the tight, European-style formations of the advancing British, easy targets as they marched forward without cover or concealment, and by the scores they fell lifeless or wounded onto the stubble of the muddy cane field. The battle quickly became a slaughter. Among the hundreds of British killed was their commander, Major General Sir Edward Pakenham, shot off his horse and mortally wounded as he attempted to rally his disintegrating army. Finally, mercifully, a retreat was effected, and the fighting ceased. Jackson and his motley American army had won a huge victory, and New Orleans had been saved, as Jackson had promised its citizens it would be.

Shreve returned to the *Enterprise* to take on additional missions for General Jackson, moving stores and the wounded from the battlefield, transporting prisoners and returning American troops to their previous posts, all of which prevented Shreve's return to Brownsville. In the meantime, Edward Livingston waited for the end of the emergency so he could make another attempt at the *Enterprise*. Actually the war had been over since December 24,

1814, when representatives of the United States and England signed the Treaty of Ghent in Belgium, but the news had not crossed the Atlantic in time to avoid the Battle of New Orleans.

On February 23 Robert Fulton died, succumbing to a series of illnesses and ailments. A little more than two months later, on May 6, Edward Livingston, now the man most interested in protecting the Mississippi River steamboat monopoly rights, sent marshals back to the *Enterprise* to seize it, and this time they succeeded.

Shreve, however, had retained another of New Orleans's ablest lawyers, A.L. Duncan, who posted bond for the boat and had it returned to Shreve's custody within hours. While Livingston was in the process of countering that move by filing a lawsuit against the *Enterprise*, Shreve fired up its boiler and shoved off into the Mississippi's current, bound for home.

The river was at flood stage then and above Natchez it was out of its banks and pouring across low-lying areas. The deeper water proved a great help to the *Enterprise*, allowing Shreve to proceed without much worry of running aground and without the bother of continually taking soundings to avoid shallow water. Making good speed, the *Enterprise* arrived in Louisville on May 31, the first steamboat to complete the voyage from New Orleans to Louisville. The *Vesuvius* had tried it and failed. The *New Orleans* and the *Comet* had not even attempted it. All three had gone *down* the Mississippi, with enough power going downstream to force their way across the sand bars that lay in the shallows, but going *up* the big river, against the current, they lacked sufficient power to drive themselves across the sand bars. Shreve guessed that the same would have been true for the *Enterprise* had it not been for the high water.

As he made the voyage from New Orleans, Shreve considered the problems of shallow water. In some ways more inventive than Fulton, who had encountered the same problems but had failed to create solutions, Shreve devised ways to overcome the problems. By so doing, he became the father of the historic Mississippi River steamboat, providing a design that was copied by most steamboat builders.

Shreve's idea was to get rid of those deep-draft hulls that made sense for sea-going vessels but were impractical for use on the Mississippi and other western rivers. His long experience with flatboats told him that the sensible way to build a riverboat was to build it with a flat, shallow hull. That meant he would not be able to mount the boiler and engine down in the hull, as Fulton and others had been doing. Instead, Shreve proposed installing the machinery on the boat's main deck. To accommodate passengers, he would build a second deck on top of the main deck. The boat would be like a floating two-story building, ungainly in appearance perhaps, but practical — and, he believed, successful in navigating shallow water.

Besides changing the fundamental design of the steamboat, Shreve wanted to give it more power. To do so, he redesigned the steam engine. Fulton's boats were powered by low-pressure, condensing engines, heavy and inefficient, with stationary, vertical cylinders. Daniel French put in his boats a high-pressure steam engine with a cylinder that oscillated on trunnions — hollow shafts through which steam was received and exhausted — as the piston, connected to a crank that rotated the paddle wheel, moved back and forth. Altering French's design, Shreve proposed using stationary, horizontal cylinders with oscillating pitmans — rods that connected the reciprocating action of the piston to the rotary action of the paddle-wheel crank. His design called for high-pressure steam that would be exhausted not into a condenser — as it was with Fulton's engine and others' — but instead would be drawn off through flues in the boiler. Eliminating the heavy, bulky condenser would, besides providing more payload space, make the boat lighter and more suitable for shallow water. The water that the condenser saved for reuse would, with Shreve's design, be replaced by river water, an inexhaustible supply of which, Shreve reasoned, would always be readily available to make steam.

When he arrived at Brownsville, having discharged the *Enterprise*'s cargo in Pittsburgh, Shreve turned the boat over to French and gave him a complete report of the voyage and the boat's performance. He told him that the boat's success in steaming up the Mississippi was owed to the high water, not the capability of the *Enterprise*. He said he had some ideas for a steamboat that would be capable of making the upriver voyage without the benefit of high water. French wasn't interested. He let Shreve know that he had confidence in his own engine and boat designs and he would stick with them. Shreve then determined to build his own steamboat, incorporating his own ideas.

The result was the *Washington*, capable of bearing four hundred tons, built in Wheeling, Virginia (now West Virginia), its engine built according to Shreve's specifications at a shop in Brownsville. The engine's cylinder would have a twenty-four-inch bore, and its piston a six-foot stroke. Work began on the *Washington* in September 1815, and by June 1816 it was ready for its maiden voyage. At least two descriptions of it survive. One is from the trade paper, *Niles' Weekly Register*, published in Baltimore:

> She is 148 feet in length. Her main cabin is sixty feet; she has three handsome private rooms, besides a commodious bar room. She is furnished in very superior style. Gentlemen from New York who have been aboard of her, assert that her accommodations exceed anything they have seen on the North [Hudson] River.... Her steam power is applied upon an entirely new principle, exceedingly

simple and light. She has no balance wheel, and her whole engine possessing the power of one hundred horses, weighs only nine thousand pounds. It is the invention of captain Shreve.[2]

The other account is from William Mercer, who was a passenger on the boat in 1816 and wrote about it in his diary:

> The boiler is placed midships on the deck, and is heated by a furnace placed at either end. The steam is conveyed through two tubes to the machinery, which is under deck in the after part of the boat, and which, being set in motion, turns a single water wheel, placed near the stern and concealed from the view of persons on the deck by a gentle elevation of the flooring timber. The arrangement below is also different. A common cabin about 80 feet long extends from the centre to either end. In the stern it opens into two apartments, one of which is a drawing room, and the other a dormitory, both appropriated, exclusively, to the use of the ladies. Towards the bow there are also two rooms, one of which is the private apartment of the captain, and in the other, the bar is kept. In the large, common room, there are 20 berths above & below, on either side of which is calculated for the accommodation of two lodgers.[3]

According to Mercer's description, the *Washington* was, unlike the boats designed by Fulton, but like French's *Comet*, a stern-wheeler. However, its single paddle wheel, as on other early stern-wheelers, did not project beyond the stern of the boat. Its aftermost edge was more or less flush with the stern. The paddle wheel was contained within the sides of the hull, making stern-wheel boats narrower and giving them some advantage over side-wheelers in straitened channels of the river.

On June 4, 1816, the brand-new *Washington*, under the command of Captain Shreve, steamed out of Wheeling, headed for New Orleans and whatever legal trouble awaited it there, passing curious onlookers who stood staring on the banks of the Ohio. The next afternoon it reached Marietta, Ohio, and remained there two days. It anchored again just below Marietta, off Point Harmor, and stayed there overnight. On the morning of June 9, while preparing to resume its voyage, the *Washington* suffered an explosion that killed thirteen persons, including crewmen and passengers, and injured several others.

The boat then began drifting without power toward the Virginia (West Virginia) side of the river and threatening to run aground when a kedge anchor was thrown overboard at the stern to stop the drift until sufficient steam pressure could be raised for the engine to power the boat. Once the pressure was raised, the crewmen were summoned aft to haul up the kedge, and while they were doing so, the end of the cylinder nearest the stern blew off, releasing a deadly stream of scalding water onto the crewmen. Captain Shreve, his mate and several others were thrown overboard by the force of the explosion. All but one of those men were rescued, although all suffered some degree of injury. The cause of the explosion, the first on western waters, was determined to be the failure of the boiler's safety valve, which had become stuck.

Shreve and his boat survived well enough for him to take the *Washington* back to port and have repairs made on it and its machinery. By early fall of 1816 the boat was ready to begin again on its maiden voyage to New Orleans. During its stop in Cincinnati Shreve's odd-looking vessel attracted a host of visitors to it, and Shreve patiently let them inspect it. He made another stop in Louisville, to take on more passengers, and on September 24 the *Washington* passed its first big test by successfully negotiating the Falls of the Ohio, where not long before, the *Enterprise*, on its second descent of the Ohio, had wrecked on the rocks. On October 7, 1816, Shreve landed the *Washington* at the wharves of New Orleans.

Edward Livingston learned of its arrival and promptly went down to the riverfront to see it and its captain. Evidently impressed with the boat's innovations, he told the thirty-year-old Shreve, "I tell you, young man, you deserve well of your country, but we shall be compelled to beat you in the courts."[4] Livingston immediately had the boat seized and held for ten thousand dollars bail.

Shreve's canny lawyer, A.L. Duncan, was prepared for that move. Refusing to let Shreve pay the bail, Duncan one-upped Livingston by asking the court to demand a ten-thousand-dollar bond from Livingston to compensate Shreve for any loss in revenue or damage he might suffer while the boat was being held in seizure, in the event Livingston, suing to assert his monopoly rights, should lose his case in court. The court granted Duncan's request. Livingston then hastily decided he didn't want to gamble ten thousand dollars and he released the *Washington* back to Shreve. A week later, with a load of passengers and cargo, Shreve headed the *Washington* back upriver.

He was prevented from reaching Louisville, however, by an early freeze that had filled the Ohio with ice that blocked the boat's passage. He docked the *Washington* at Shippingport, Kentucky, below the falls, about two miles from Louisville, and left it there to await the spring thaw while he waited in Louisville, close to his ice-trapped steamboat. He sent for his wife, Mary, and their children to join him in Louisville, where he would later establish his residence. It was shortly after they arrived that the Shreves suffered the death of their baby son, Zane.

When the ice in the Ohio broke up, Shreve was ready to restart operations, and on March 3, 1817, the *Washington* shoved off with freight and passengers, bound for New Orleans.

At New Orleans, which the *Washington* reached on the night of March

12, the legal battle resumed. Livingston again had the boat seized, and Duncan again petitioned to require Livingston to post a bond against possible loss. Livingston argued against the bond, but the court ruled against him. Livingston returned custody of the boat to Shreve and then huddled with his staff of lawyers. From that conference came a new tactic. Seeing that Shreve and Duncan would not be intimidated into giving up, Livingston decided on a move that was very much like a bribe. The monopoly would press its lawsuit against Shreve and the *Washington* in federal court, while its suit against the *Enterprise* still languished in the Louisiana appellate court, and the monopoly holders would offer Shreve a half interest in their monopoly rights on condition that Shreve would arrange with Duncan to lose the case in federal court, thus protecting the monopoly. It was a shrewd maneuver, since bringing Shreve into the monopoly's operations would not only keep the monopoly rights intact but would bring Shreve and his boat into their business, providing the monopoly holders with a boat that could reliably steam upriver beyond Natchez, something they did not have.

It was a tempting offer, representing a windfall to Shreve and the end of his legal hassles as well. But, more concerned with a free Mississippi, open to all comers, than a good deal for himself, Shreve turned the offer down.

Through the federal court Livingston quickly struck back at Shreve's refusal. Shreve was arrested and held on ten thousand dollars bail and ordered to appear before the court on the third Monday of April to answer Livingston's complaint. After twenty-four hours in jail Shreve was freed and he quickly steamed off aboard the *Washington* on March 25, just two days behind its scheduled departure.

On April 21, 1817, a hearing was held in the District Court of the United States for the Louisiana District, presided over by Judge Dominick A. Hall. Shreve as well as the monopoly holders were represented by their attorneys. The court record tells the disposition of Livingston's case against Shreve:

> It appearing, after arguments of counsel and the examination of the record in the case, that the court has no jurisdiction of the same, it is therefore ordered, adjudged and decreed that the petition of Pltffs [plaintiffs Livingston and his wife, Elizabeth] be dismissed with costs.[5]

Edward Livingston and the monopoly had been defeated again. Even so, Livingston clung to the monopoly owners' claim of exclusive rights to steamboat navigation at the port of New Orleans. Shreve meanwhile made regular trips to and from New Orleans, not only with the *Washington* (which had made the run to Louisville in twenty-one days, less than a quarter of the time taken by keelboats and barges), but with new steamboats that he and a set of partners built—the *Ohio*, built in 1817, the *Napoleon*, built in 1818, and the

Post Boy, built in 1819. Whatever hopes Livingston held while he waited for a decision from the Louisiana State supreme court, where his appeal in the *Enterprise* case was bottled up, were finally exhausted in 1819, and the company that had succeeded to Fulton and Livingston's steamboat monopoly on the Mississippi at last withdrew all claims to an exclusive right to operate steamboats on the Mississippi.

Persistent, determined, unafraid, Henry Shreve had won his fight — and not just for himself. Judge Samuel Treat, later writing about the historic victory in a nineteenth-century magazine article, "Political Portraits With Pen and Pencil," remarked, "At this day, the enthusiasm with which the news was received cannot be duly appreciated ... the western country owes a vast debt to Captain H.M. Shreve."[6]

Shreve had broken the stifling monopoly and freed the Mississippi for the entrepreneurial spirit of the growing nation.

The *Washington* continued to make round trips between New Orleans and Pittsburgh until it became worn out and obsolete and was scrapped in 1822. Shreve never stopped trying to improve on it. He figured out a way to eliminate the main disadvantage of side-wheelers, notorious for their wide turning radius. He connected a separate engine to each of the two paddle wheels, so that one wheel could be reversed while the other rotated forward, allowing a boat to turn about within its own length. He also decided that additional decks could be stacked atop the second deck he had already introduced, and when he built the *George Washington* in 1824, he added two decks to provide more cabins — making it a four-story structure, with a pilothouse atop it — and a promenade to give passengers more room to move about. A passenger named Bullock voyaged on the *George Washington* two years after it was put into service and described the vessel:

> On the third of April we left New Orleans in the beautiful steam-boat George Washington of 375 tons, built in Cincinnati, and certainly the finest fresh-water vessel I have ever seen.... The accommodations are excellent, and the cabins furnished in the most superb manner. None of the sleeping rooms have more than two beds. The principal [rooms] are on the upper story, and a gallery and verandah extends entirely round the vessel, affording ample space for exercise, sheltered from the sun and rain, and commanding from its height, a fine view of the surrounding scenery, without being incommoded by the noise of the crew passing overhead. The meals served ... are excellent, and served in superior style. The ladies have a separate cabin, with female attendants, and laundresses; there are, also, a circulating library, a smoking and drinking room for the gentlemen, with numerous offices for the servants &c. &c....[7]

The *George Washington* had set the standard for riverboats, and not just those on the Mississippi, but on rivers everywhere. Shreve made the Mississippi River steamboat an American institution and in so doing played a huge part in the development of the nation. "To him," the *St. Louis Republican* declared when it published his obituary following his death in 1851, "belongs the honor of demonstrating the practicability of navigating the Mississippi with steamboats."

• 7 •

The Proliferation

By 1815, just eight years after the *North River Steam Boat* had made its historic voyage up the Hudson, the Fulton-Livingston company had built twenty-one steamboats, all of them designed by Fulton. Master of the Hudson by virtue of the monopoly granted it by the New York State legislature, the firm was reaping profits from five steamboats in service on the Hudson, including a new version of the *North River*, rebuilt from the original and nearly twice as big, the *Car of Neptune*, the *Paragon*, the *Richmond* and the *Chancellor Livingston*, which was not completed until after Fulton's death. On the Mississippi the Fulton-Livingston company had similarly expanded, operating the *New Orleans* (until 1814, when it sank), the *Aetna*, the *Natchez* and the *Buffalo*.

It operated five steam ferryboats from Manhattan — the *Firefly*, the *Jersey*, the *York*, the *Camden* and the *Nassau* — and two ferries that plied Long Island Sound, the *Connecticut* and the *Fulton*. In addition, it operated the *Washington* on the Potomac River and the *Olive Branch* and the *Raritan* on the Raritan River in New Jersey. In an attempt to establish a steamboat monopoly in Russia, the firm had built the *Empress of Russia*, and winning himself one more distinction, Fulton had also designed the world's first steam warship, the *Demologos* — which the U.S. Navy called *Fulton the First* — the guns of which boomed a salute to Fulton in New York's harbor on the day of his funeral.

On March 2, 1824, in a ruling written by Chief Justice John Marshall, deciding the case of *Gibbons v. Ogden*, the New York law that gave Fulton and Livingston their steamboat monopoly was struck down by the Supreme Court of the United States. The Hudson, like the Mississippi, like every other navigable river and lake in the nation, was open to all comers, all vessels, however propelled. It was a grand new day for steamboats in America.

The statistics indicate the difference that *Gibbons v. Ogden* made. In 1819 there were eight steamboats operating on the Hudson River. In 1826, two years after the Supreme Court's ruling, there were sixteen Hudson River steamboats; in the late 1830s there were forty-five, and by 1840 there were more than one hundred. More than a dozen steamboat companies were established to operate in New York. In 1849 passengers could choose between twenty steamboats that ran daily between New York City and Albany. Cornelius Vanderbilt became the owner of more than fifty steamers, operating on several routes, and began amassing the fortune that would later make him the richest man in America.

Steamboats were also multiplying elsewhere in the Northeast. Lake Champlain, gateway for trade between Canada and New York State, was the second oldest water route regularly traveled by steamboats, its first steamer being the *Vermont*, built in 1808 at Burlington, Vermont, on the lakeshore. By July 1821 a Lake Champlain excursion-boat service had been established, using the steamer *Congress* to carry, as its advertisement read, "Parties of Pleasure, and others, who may wish to view the remains of those ancient fortresses, Ticonderoga and Crown Point, and other more recently memorable places on the Lake, such as the Battle Ground of Macdonough's Naval Engagement — Plattsburgh, &c."[1] The *Congress* steamed out of Whitehall every Thursday morning at five o'clock. Excursion passengers disembarked from it on the second day and boarded the southbound steamer *Phoenix* for a return trip to Whitehall while the *Congress* continued north to Canada.

By 1842 at least sixteen steamboats had been put into service on Lake Champlain. On one of them, the *Burlington*, the renowned British novelist Charles Dickens traveled as a passenger in 1842 and he wrote fulsomely about the experience:

> There is an American boat — the vessel which carried us on Lake Champlain, from St. John's to Whitehall, which I praise very highly, but no more than it deserves, when I say that it is superior to any other in the world. The steamboat, which is called the *Burlington*, is a perfectly exquisite achievement of neatness, elegance and order. The decks are drawing-rooms; the cabins are boudoirs, choicely furnished and adorned with prints, pictures and musical instruments; every nook and corner in the vessel is a perfect curiosity of graceful comfort and beautiful convenience.[2]

Steamboat service blossomed along the Connecticut coast. The *Lafayette* was built in 1828 to operate out of Bridgeport, and in 1835 one of the Vanderbilt steamboats, the *Nimrod*, was shifted from the Hudson River to Bridgeport. Thanks largely to the building of railroads that terminated there, Bridgeport in the 1840s became a busy port for steamers, which brought freight from New York City to the Bridgeport railroad terminals for shipment by rail to the interior of Connecticut and New York State.

7 • *The Proliferation* 91

New Haven was introduced to steamboats when the *Fulton*, the last boat built under Fulton's supervision and which he designed specifically for service on the waters of Long Island Sound, arrived there from New York in March 1815. It left New York a little after five o'clock on a Tuesday morning and landed at New Haven at four-thirty that afternoon, which was not considered a speedy trip. But the New York *Evening Post* reporter who told of its introductory voyage speculated that when hindering mechanical problems were solved and the weather was good, the trip would be made in eight or nine hours. He had other good things to say about the boat. "We believe it may be affirmed," he wrote, "that there is not in the whole world such accommodations as *Fulton* affords. Indeed it is hardly possible to conceive that anything of the kind can exceed her in elegance and convenience."[3]

The New Haven Steamboat Company, formed in 1822, operated the *United States* and the *Hudson* out of New Haven. The *United States* was said to be the first steamboat with a pilot house.

Steamboats were serving customers on Nantucket Sound, too, running between New Bedford and Edgartown and providing service to Nantucket, Martha's Vineyard and Hyannis. Steamers, including the *Massachusetts*, the *Connecticut*, the *Fanny* and the *Merrimack*, operated from other Massachusetts ports as well. The *Tom Thumb*, thirty feet in length, was the first steamboat to appear in Maine, where after being towed from Boston, it steamed up the Kennebeck River in 1818. It was later joined in service in, to and from Maine by the *Kennebeck*, the *Patent*, the *Maine*, the *Waterville*, the *Legislator* and the *New York*, among others.

On the Delaware River, Robert Livingston's brother-in-law, John Stevens, first operated the *Phoenix*, then replaced it with the *Philadelphia* in 1815. Two other steamers, the *Bristol* and the *Sea Horse*, later entered the competition on the run between Philadelphia and Bristol, Pennsylvania. Stevens's company, the Union Line, added the *Rain Bow*, the *Swan*, the *Stevens*, the *Stockton*, the *Nelson*, the *Burlington*, the *Trenton* and the *Belknap* to the steamboats serving the upper Delaware. Others followed.

Steamboats also ran between Philadelphia and Wilmington, Delaware, the first of which, the *Vesta*, was in service by 1820. After it came the *Superior* and the *Wilmington*. Steamer service from Philadelphia to Salem, New Jersey, below Wilmington, began in 1824 and was initiated by the *Lafayette*, followed by the *Albemarle*, the *Essex*, the *Proprietor*, the *Linnaeus*, the *Flushing* and the *Pioneer*, among others. Cape May, New Jersey first received steamboat service in 1824, the earliest steamers including the *Delaware*, the *Ohio* and the *Robert Morris*.

Baltimore's first steamboat was the *Chesapeake*, built in Baltimore in 1813 at a cost of forty thousand dollars. It soon had to compete with the *Eagle* and after that came the *Virginia*, the *Norfolk*, the *Roanoke*, the *Surprise*, and the *Richmond* and others later, running between Baltimore and Norfolk and Richmond, Virginia. When the marquis de Lafayette, the French general who had aided the American cause in the Revolutionary War, came back to visit the United States in 1824, there were five steamboats gathered in Baltimore harbor to extend an official greeting and welcome him back to the grateful nation whose independence he had helped win.

Steamboats were also plying the Great Lakes in the 1820s, one of the earliest being the *Walk-in-the-Water*, which was launched into Lake Ontario in 1819. By 1826 there were seven steamers operating on the lakes. In 1833 there were eleven steamboats serving Buffalo, New York, and together they carried more than sixty thousand passengers to and from Buffalo.

By 1835 steamboats were also navigating the waters off the Atlantic Coast, running between New York and Charleston, South Carolina. Those boats included the *David Brown*, the *William Gibbons*, the *Columbia* and the *New York*. In 1835 the *Columbia*, owned by Charles Morgan of New York, became the first steamboat to operate in the Gulf of Mexico, running between New Orleans and Galveston, Texas. Morgan's sea-going, iron-hulled steamers also established service between New York and New Orleans and ports on the Mexican coast, and ran steamboats from New Orleans on Lake Pontchartrain, through Lake Borgne to Mobile on the gulf.

In North Carolina, steamboats, including the *Prometheus*, operated from the mouth of the Cape Fear River up to Wilmington; and in South Carolina, steamboats regularly plied the lower Waccamaw River and the Ashley. Steamboats appeared in Georgia as early as 1816, operating on the Savannah River, eventually running between Savannah and Augusta, and in 1828 they began making regular runs on the Georgia river system that includes the Apalachicola, Chattahoochie and Flint. In 1829 the first steamer arrived at Macon, on the Ocmulgee River, and in 1833 a commercial steamboat service was begun between Darien, on the Georgia coast, and Macon.

In Alabama the steamboat era began when the steamer *Alabama*, built at St. Stephens, Alabama, on the Tombigbee River, upstream of Mobile, was launched in 1818. Its engine, though, lacked sufficient power to take it back up the river after it had gone with the current down to Mobile. The *Alabama* was followed by the more powerful *Mobile*, which made it up the Tombigbee and Black Warrior rivers all the way to Tuscaloosa. The next steamboat to operate out of Mobile was the *Harriet*, which successfully ran from Mobile to Montgomery. Cotton was the mainstay of Alabama's economy then, and the Alabama River, winding through the southern half of the state, carried

7 • *The Proliferation*

it, aboard the multiplying steamboats, to Mobile where it could be shipped across the gulf to far-flung destinations.

The greatest growth in steamboat numbers, however, was on the Mississippi and its tributaries. Between 1811, when Nicholas Roosevelt set out for New Orleans aboard the *New Orleans*, and 1820 at least sixty steamboats were either built on the western rivers or sent to them to begin operations. An early twentieth-century record[4] shows the increase and the total steamboats in service and compares the western-water numbers with those elsewhere. It also reveals the increasing size of the vessels.

Year	Number of Steamboats on the Mississippi	Approximate Tonnage
1834	231	39,000
1840	225	49,000
1843	672	134,400
1844	686	144,150
1845	789	157,713
1847	958	200,000
1849	1,000 (probable)	250,000 (probable)

Steamboat tonnage by sections and cities of the United States, operating in 1842.

Southwest:

New Orleans	80,993
St. Louis	14,725
Cincinnati	12,025
Pittsburgh	10,107
Louisville	4,618
Nashville	3,810

Total Southwest 126,278

Northwest:

Buffalo	8,212
Detroit	3,296
Presque Isle	2,315
Oswego	1,970
Cuyahoga	1,859

Total Northwest 17,652

Seaboard·

New York	35,260
Baltimore	7,143
Mobile	6,982
Philadelphia	4,578
Charleston	3,289

Newbern	2,854
Perth Amboy	2,606
Apalachicola	1,418
Norfolk	1,395
Boston	1,362
Wilmington	1,212
Georgetown	1,178
Newark	1,120
Miscellaneous	4,767
Total Seaboard	76,064
Grand Total	219,994

Those statistics show that in 1842 steamboat tonnage on the Mississippi and its tributaries accounted for more than 57 percent of the total steamboat tonnage of the nation. In 1842 Pittsburgh alone had more steamboat tonnage than Philadelphia, Boston and Charleston combined. The tonnage of steamboats operating out of Pittsburgh and Cincinnati together exceeded the total tonnage of all Great Lakes ports combined by five thousand tons. In 1842 the steamboat tonnage on the Mississippi and its tributaries exceeded by forty thousand tons the total 1834 tonnage of steamers in England, Scotland, Ireland and the then-British dependencies combined. As far as steamboats were concerned, by the middle of the nineteenth century, the Mississippi and its tributaries were definitely where the action was.

All those steamboats represented a busy and growing boat-building industry. Cincinnati, which led the towns of the Ohio River valley in the number of steamboats built, in 1843 employed some seven hundred and seventy persons in boat-building, constituting a significant proportion of its then meager population. The statistics clearly show the steamboat's growth. In 1820 the Ohio valley boat-building towns built fifteen steamboats, with a total tonnage of 2,643 tons. In 1830 those towns built thirty-three boats, totaling 4,881 tons. In 1840 they built sixty-three steamboats, totaling 9,224 tons. In 1850 they built one hundred and nine steamboats, with a total tonnage of 20,911 tons.

In 1843, a single year, Cincinnati alone produced forty-five steamboats, with a total tonnage of 12,035 tons; Pittsburgh produced twenty-five, totaling 4,347 tons; and the clustered towns of Louisville, Kentucky; New Albany, Indiana; and Jeffersonville, Indiana, together produced thirty-five steamboats, totaling 7,406 tons. Boatyards along the Ohio River were hard-pressed to keep up with the demand. They put on extra work crews that worked at night by torchlight for double-time wages. Machine shops and foundries as well toiled furiously to keep up with orders.

Some of those vessels did not add to the size of the steamboat fleet but were replacements for boats that wrecked or burned or were otherwise lost or those that simply wore out after several years of service, the average lifespan of a steamboat then being only about five years. Steamboat owners who lost their vessels to accidents or obsolescence usually were quick to replace them. So eager were owners to keep their steamboat business humming, and the cash returns flowing, that they often would order new vessels from the boatyards within twenty-four hours of having lost a boat to a snag, fire or explosion.

The U. S. Treasury Department reported that in 1842 steamboat tonnage on the Mississippi and its tributaries amounted to 70,033 tons, and in 1851 that tonnage doubled. The rise in total tonnage during the second quarter of the nineteenth century of course came from increases in size and carrying capacity as well as in the number of steamboats. The following table shows the increase in capacity.[5]

Years	Average Capacity in Tons (Downstream)	Average Capacity in Tons (Upstream)
Before 1820	110	55
1820–1829	232	116
1830–1839	310	155
1840–1849	496	248
1850–1859	630	315

Through improvements in construction techniques during the second quarter of the nineteenth century, Mississippi steamboats, built almost entirely on the Ohio, not only got bigger but much more efficient as payload carriers, boosting the amount of freight they could carry in proportion to their size, more than tripling the ratio of their carrying capacity to their tonnage, from 0.50 to 1.75. That increase, pure joy to steamboat owners, meant that a boat of two hundred tons, for example, built before the 1820s, could carry no more than 100 tons of freight, but a steamboat of two hundred tons built in the 1850s could carry 350 tons.

Paddle wheels on sidewheelers and sternwheelers alike evolved over the same period, becoming larger in diameter, thereby increasing the speed of the boat without significantly increasing the amount of fuel consumed.

Another important technological advancement in steamboat construction was the use of high-pressure engines, which were developed more or less through trial and error over a period of years on the western rivers. The high-pressure engine presented a number of advantages over the low-pressure, condensing engine. For one thing, important in the relatively shallow Mississippi and other western waters, it was some 60 percent lighter than a comparable low-pressure engine, making the steamboat lighter and its draft shallower.

Not only was the high-pressure engine more powerful than its low-pressure counterpart, it was also a simpler machine, simpler to manufacture, and thus it cost steamboat owners about 60 percent less than the more complicated low-pressure engine. It was more easily maintained and repaired, usually requiring no more skilled a mechanic than was the boat's engineer — a big advantage when mechanical trouble occurred on the river, miles from a machine shop. It was also less susceptible to problems arising from the boats' use of silt-laden river water to make steam.

Steamboat designers soon figured out that longer boats made better boats, and the marked increase in tonnage was a logical result of that conclusion. A longer hull increased the boat's buoyancy and speed, giving it a shallower draft and making it faster than a boat of similar tonnage but with a shorter hull.

By the late 1830s the design of Mississippi steamboats had become settled and remained standard for years to come, even for boats built in the late twentieth century. They were shallow-draft, flat-bottomed, multi-decked vessels with tall, twin chimneys, or smoke stacks. There was no mistaking them for anything else when they appeared, riding majestically upon the waters of the Mississippi.

Some steamboat owners, out to get as much service as possible from their investment, operated boats smaller than the average Mississippi steamboat, which in 1851 was two hundred and seventy-five tons. Those smaller boats were light enough to run on the Mississippi and Ohio during dry periods when the water was low and on small tributary streams when the water there was high.

A typical Mississippi sternwheeler at that time would be two hundred feet long and thirty feet in the beam and would measure around two hundred tons. Its carrying capacity would be three hundred tons of freight, and its passenger capacity would be about two hundred persons in its staterooms and about one hundred on its main deck. The tables below give an idea of the revenue a typical Mississippi steamboat produced from the freight and passengers it carried. They are the actual rates[6] on the upper Mississippi in 1857.[7]

Freight rates per ton, going upstream—

Less than 30 miles	6 cents per mile
30 to 60 miles	5 cents per mile
More than 60 miles	4 cents per mile

Freight rates per ton, going downstream—

Less than 30 miles	5 cents per mile
30 to 60 miles	4 cents per mile
More than 60 miles	3 cents per mile

7 • The Proliferation

Passenger rates, going upstream—

Dunleith, Illinois, or Galena, Illinois, to:	Miles	Cabin Fare	Deck Fare
Cassville	30	$ 2.00	$1.25
Prairie du Chien	66	3.50	2.00
La Crosse	150	6.00	3.25
Red Wing	256	10.00	5.00
St. Paul and Stillwater	321	12.00	6.00

Prairie du Chien, Wisconsin, to:

St. Paul, Minnesota	255	10.00	5.00

La Crosse, Wisconsin, to:

St. Paul, Minnesota	175	7.00	4.00

Passenger rates, going downstream—

St. Paul or Stillwater to:

Hastings	32	$ 1.50	$1.00
Red Wing	65	2.50	2.00
Winona	146	4.50	2.50
La Crosse	175	5.00	3.00
Prairie du Chien	255	7.00	3.50
Dunleith or Galena	321	8.00	4.00

    The average upstream trip produced total receipts estimated at $4,450. Of that total, $3,000 came from freight. An average of a hundred and fifty cabin passengers, paying an average fare of eight dollars each, yielded a total of $1,200, and an average of fifty deck passengers, paying an average fare of five dollars each, yielded $250, for a grand total from passenger fares of $1,450.

    Downstream trips were not so profitable, the big reason being that there was less freight to carry. The only significant cargo going downstream from St. Paul was wheat, which was shipped in two-bushel sacks, each weighing one hundred and twenty pounds. Boats going downstream carried an average of five thousand sacks of wheat at an average rate of 12 cents per sack, making their total freight receipts $600. They also carried an average of eighty passengers at an average fare of eight dollars each, making a total of $640 in fare receipts and bringing total receipts for the downstream trip to $1,240.

    The round trip therefore produced a total of $5,690 on average. During the five-month season for navigating the upper Mississippi a steamboat would make four round trips a month, earning an average of $22,760 per month. Crew salaries and wages totaled an estimated $5,850 for a month. Food supplies, figured at $75 a day for thirty days, totaled $2,250. Wood for fuel, estimated at twenty-five cords per day at $2.50 per cord for thirty days totaled $2,000, and miscellaneous other expenses were estimated at $1,400

The steamer *Imperial* arrives at New Orleans on July 16, 1863, reopening the Mississippi River to unimpeded passage during the Civil War. The *Imperial* left St. Louis on July 8, four days after Union troops had captured the strategic Mississippi River city of Vicksburg and eliminated the last major barrier to free commerce on the river (Library of Congress).

for a month. The monthly total for all expenses, not counting depreciation, was $11,500.

And so for the five months that the average boat operated during the year on the upper Mississippi, it returned to its owner $56,300 in profits. During the middle of the nineteenth century, an average Mississippi River steamboat cost between $20,000 and $50,000 to build or buy, substantially less than a year's profits from its operation. A boat would more than pay for itself in its first year of service.

On the lower Mississippi cotton was the predominant cargo. A steamboat of three hundred and fifty tons, which might cost as much as $50,000, could carry five hundred tons of freight, or some fifteen hundred bales of cotton, stacked high on its decks. Cotton generally was shipped to New Orleans from the two major ports of the Mississippi valley's cotton-growing areas — Memphis and Natchez. In 1846 the shipping charge for cotton from Memphis to New Orleans was $2 a bale; from Natchez, $1 a bale. On the return trip, the freight rate from New Orleans to Natchez was 75 cents per one hundred pounds, more to ports more distant, on a steamboat that could carry from the wharves and warehouses of New Orleans five hundred tons of everything needed or wanted by the people of America's burgeoning interior.[8]

Doing the arithmetic on those numbers shows just how lucrative Mississippi River steamboating was for boat owners and helps explain the proliferation of steamboats on western waters.

It was the nation's westward expansion, however, that was the real and irrepressible force driving the demand for transportation and increasing commerce on the big river and its tributaries. The population of the United States grew from 5,306,000 in 1800 to 23,192,000 in 1850, or 33 percent every decade. West of the Alleghenies and Appalachians, in the valleys of the Mississippi's tributaries, the population increase was even more remarkable. From a total of 560,000 in 1800 the population there swelled to 10,520,000 in 1850, an average increase of 182 percent in every decade in the first half of the nineteenth century, during the same period that steamboats were proliferating on the Mississippi and its tributaries.

The Mississippi steamboat both served and helped cause that population growth. It opened up the middle of the country for settlement and it brought in the new immigrants who would do the settling. The population increase in the major ports along the Ohio and Mississippi gave evidence of the steamboat's role in the western expansion. In the years between 1820 and 1850 the population of Pittsburgh increased from 4,700 to 46,000. Cincinnati's population jumped from 9,600 to 115,000, Louisville's from 4,000 to 43,000, and St. Louis's from 5,000 to 77,000.

In 1830, New Orleans, yet to experience the full effect wrought by the steamboats that came to its wharves, had a population of 46,000. Ten years later, in 1840, with steamboat commerce surging, New Orleans had become the third city in the nation, after New York and Baltimore, to reach 100,000 population.

By the middle of the nineteenth century, steamboats on the Mississippi had introduced into American history a whole new age. And there was more to come.

PART THREE. THE CIRCUMSTANCES

• 8 •

The Sweet Life on the Mississippi

The man first boarded the steamboat at Natchez and took one of the vessel's best cabins. When the boat arrived in New Orleans, he notified its officers that he would not be disembarking but would keep his cabin and stay on the boat on its return trip to St. Louis. When it reached St. Louis, he again declined to get off and made arrangements to take it back to New Orleans. At New Orleans, he bought passage back to St. Louis. For two months he stayed on the steamer as it voyaged between New Orleans and St. Louis, which raised vexing questions in the captain's mind. Elderly and friendly, the man jovially mingled with the other passengers, warmly greeting them, offering them cigars, buying them drinks, sitting with them and talking and apparently enjoying their company. The captain decided the mysterious passenger was not a gambler, to whom destinations were also unimportant. Nor was he, as passengers occasionally were found to be, a thief or a murderer hiding out from the law. He just didn't seem that sort. The captain couldn't stop wondering about the man.

At last, with as much tact as he could manage, he asked the man if he would mind telling him why he was making the repeated trips.

"Of course, sir, I'll tell you," the old gentleman answered. "It's the finest way to pleasure myself that I know. No hotel in America can equal this. The finest food—your wild game, your glazed fish, your roasts, sauces and pastry! My cabin—it's as finely equipped, as well decorated, as any room I've enjoyed in my life. The bar, the cabin, the promenade—nothing to match them, I tell you. And the company! I meet all my friends, the best people in the world. Why should I want to leave?"

Beaming with satisfaction, the captain treated his happy passenger to a drink at the bar.

That is a story out of Mississippi River steamboat lore which has in it enough truth to make it believable. Samuel Clemens, whose charming powers of observation of his fellow humans, augmented by his experience as a Mississippi River steamboat pilot, provided him broad knowledge of steamboats and their passengers, was among others who agreed with that traveler. People, Clemens wrote, compared Mississippi River steamboats to other things they had seen, "and, thus measured, thus judged, the boats were magnificent.... The steamboats were finer than anything on shore. Compared with superior dwelling-houses and first-class hotels in the [Mississippi] valley, they were indubitably magnificent, they were 'palaces.' To a few people living in New Orleans and St. Louis they were not magnificent, perhaps; not palaces; but to the great majority of those populations, and to the entire populations spread over both banks between Baton Rouge and St. Louis, they were palaces; they tallied with the citizen's dream of what magnificence was, and satisfied it."[1]

Clemens' intimate knowledge of Mississippi River steamboats dated back at least to 1856, when he decided he would become a pilot, a job that he officially began when he got his pilot's license in 1858 and that ended when he quit it at the beginning of the Civil War in 1861. He described the Mississippi River steamboat as he knew it:

> When he [the passenger] stepped aboard a big fine steamboat, he entered a new and marvelous world: chimney-tops cut to counterfeit a spraying crown of plumes — and maybe painted red; a pilot-house, hurricane-deck, boiler-deck guards, all garnished with white wooden filigree-work of fanciful patterns; gilt acorns topping the derricks; gilt deer-horns over the big bell; gaudy symbolical picture on the paddle-box, possibly; big roomy boiler-deck, painted blue, and furnished with Windsor arm-chairs; inside, a far-receding snow-white "cabin"; porcelain knob and oil-picture on every stateroom door; curving patterns of filigree-work touched up with gilding, stretching overhead all down the converging vista; big chandeliers every little way, rainbow-light falling everywhere from the colored glazing of the skylights; the whole a long-drawn, resplendent tunnel, a bewildering and soul-satisfying spectacle! In the ladies' cabin a pink and white Wilton carpet, as soft as mush, and glorified with a ravishing pattern of gigantic flowers. Then the Bridal Chamber ... whose pretentious flummery was necessarily overawing.... Every stateroom had its couple of cozy clean bunks, and perhaps a looking-glass and a snug closet; and sometimes there was even a wash-bowl and pitcher, and part of a towel which could be told from mosquito-netting by an expert — though generally these things were absent, and the shirt-sleeved passengers cleansed themselves at a long row of stationary bowls in the barber shop, where were also public towels, public combs, and public soap.
>
> Take the steamboat which I have just described, and you have her in her highest and finest, and most pleasing, and comfortable, and satisfactory estate....[2]

Another who knew the Mississippi River steamboat well, writer Arthur E. Hopkins, recorded his description:

> Steamboating had a romance and glamor never attained in any other kind of transportation. The large sidewheel passenger steamboat was beautiful. Her lines,

with a graceful sheer, made her set on the water like a swan; the ornamental railings were filigree of woodwork; her smokestacks towered high above the water line and their tops were cut to represent plumes or fern leaves. From the hull to the hurricane deck the boat was painted a glistening white, with the tops of the wheelhouses a sky blue, as was the breeching around the smokestacks. The pilothouse with its ornamental crown added to the appearance of the entire structure. The dome of the pilothouse matched in color the wheelhouse. A red line near the top of the hull extended from the stem to the stern, and the skylights or ventilators over the main salon were of stained glass. The main cabin, which extended nearly the full length of the boat, was done in white and gold; the walnut or rosewood of the panels at the stateroom doors provided an agreeable contrast.

There was usually a small landscape over the stateroom doors.... The bridal suites and the ladies' cabins were models of decoration; French plate mirrors in hand-carved and gilded frames adorned them; marble-topped tables, deeply velveted upholstered chairs and settees were provided; and a piano of the best make completed the furnishings.

The name of the boat painted on the sides of the wheelhouses was a triumph of the sign painter's art; it was frequently done in gold leaf. Sometimes immediately above the name of the boat was painted a landscape or figure. The boat's colors were beautiful. Flying from the forward flagpole, called the jackstaff, was a long flag outlined in red, white, and blue, with the name of the boat in red on white ground.... Inboard on each wheelhouse was a flagstaff which flew burgees bearing the names of the cities between which the boat operated. On the flagstaff at the rear of the texas the union jack was flown and on the rear flagstaff, called the verge-staff, flew the Stars and Stripes....[3]

"In the middle of the nineteenth century," another veteran steamboatman remarked, "many an artist whose canvases found no market in the older cities, found ready bidders for his brush to decorate the thirty-foot paddle boxes of the big side-wheelers with figures of heroic size." The paddle-boxes of the *Minnesota Belle*, he observed, "were decorated with pictures the same on each side, representing a beautiful girl, modestly and becomingly clothed, and carrying in her arms a bundle of wheat ten or twelve feet long, which she apparently had just reaped from some Minnesota field.... The *Northern Belle* also had a very good looking young woman upon her paddle-boxes. Evidently she exhibited herself out of pure self-satisfaction, for she had no sheaf of wheat, or any other evidence of occupation. She was pretty, and she knew it."[4]

All the steamers with "Eagle" in their names seemed to have a huge eagle embellishing their paddle-boxes; the steamer *Minnesota* bore a reproduction of the state's coat of arms; boats named for noted persons tended to reproduce a likeness of their namesakes on their paddle-boxes. But most sidewheelers, according to one account, offered paddle-box decoration no more original than a sunburst, outside of which, along the curved edge of the wheel's housing, was painted the name of the line or company that owned the boat.

Frederick Law Olmsted, the nineteenth-century landscape architect who made a name for himself not only as a creator of New York's Central Park but as a travel writer, described less grand, smaller steamboats, many of which operated on the Mississippi's tributaries:

> They are but scows in build, perfectly flat, with pointed stem and square stern. Behind is one small wheel, moved by two small engines of the simplest and cheapest construction. Drawing but a foot of water they keep afloat in the lowest stages of the rivers. Their freight, wood, machinery, hands and steerage passengers are all on the main deck. Eight or ten feet above, supported by light stanchions in the floor used by passengers, one long saloon 8 or 10 feet wide which stretches from the stern to the smoke pipes far forward.
>
> The saloon is lined on each side with staterooms, which also open out upon a narrow upper gallery. Perched above all this is the pilot house, and a range of staterooms for the officers, pilots and visiting pilots, popularly known as "Texas." Inveterate card players retire to this "Texas" on Sundays when custom forbids cards in the saloon. A few feet of the saloon are cut off by folding doors for a ladies' cabin. Forward of the saloon the upper deck extends around the smoke pipes, forming an open space, sheltered by the pilot deck and used for baggage and open-air seats.
>
> Such is the contrivance for making use of their natural highways. And really admirable it is, spite of the drawbacks, for its purpose. Roads in countries so sparsely settled are impractical. These craft paddle about, at some state of water, to almost everyman's door, bringing him foreign luxuries, and taking away his own productions.[5]

By 1859, during the Mississippi steamboat's heyday, there were thirty-two elegant passenger steamboats operating between New Orleans and St. Louis. The steamer *Eclipse*, in the years before the Civil War, was widely considered the most outstanding of them all, in size, speed and luxury. In 1853 it made the 1,440-mile trip from New Orleans to Louisville in four days, nine hours and thirty-one minutes, a record-breaking time made all the more remarkable because of the vessel's size. It measured 363 feet long and 36 feet in the beam. (In comparison, the modern Mississippi River excursion steamboat *Mississippi Queen* measures 382 feet in length and 68 feet in the beam.) *Eclipse*'s saloon glistened with gilt and was adorned with rich, colorful paintings. The saloon was divided roughly in half, according to gender, and at the men's end stood a gilt statuette of Andrew Jackson, and at the women's end stood a matching statuette of Henry Clay. Included among the lavish furnishings was a piano for the use of passengers. Special sleeping rooms were available for the passengers' servants. The boat could accommodate as many as 180 passengers, along with its 121-member crew. Most extraordinary of all its attractions, the *Eclipse*'s accommodations included no less than forty-eight bridal chambers, a telling testimony to the steamboat's power to inspire romance in the bosoms of its passengers. Other usual amenities aboard steamboats included a post office, a laundry and a library.

8 • The Sweet Life on the Mississippi

A steamer's main cabin at dinnertime. Also called the grand saloon, the main cabin, located on the boiler deck, served at various times as a sumptuous hotel lobby, a lounge, a dining room, a ballroom or a concert hall. For first-class passengers the grand saloon was the magnificent great hall of a wondrously beautiful floating palace, illuminated by glistening cut-glass chandeliers, decorated with oil paintings and thick carpets. At dinnertime "steaming foods [were] piled high on the long linen table cloth," one passenger reported, "...with attentive waiters standing at the traveler's elbow, waiting with more food ... neither homes nor hotels of the [eighteen] fifties were ever like this" (Library of Congress).

To a great many, a voyage as a passenger on a Mississippi River steamboat was, as one writer of the early twentieth century called it, "a luxurious orgy." The grand saloon was more comfortable, more ornate, more sensuous than the parlors or sitting rooms of the passengers' homes, which, in the custom of their times, they entered and used only on special occasions. For those passengers the Mississippi River steamboat's saloon was the magnificent great hall of a wondrously beautiful floating palace. "The wooden filigrees that stretched down the long aisle in a tapering vista illuminated by the glistening cut-glass chandeliers; the soft oil paintings on every stateroom door; the thick carpets that transformed walking into a royal march; the steaming foods piled high on the long linen cloth in the dining room, with attentive waiters standing at the traveler's elbow, waiting with more food, and gaily colored desserts in the offing — neither homes nor hotels of the [eighteen] fifties were ever like this."[6] At various times the grand saloon could be a sumptuous hotel lobby, a lounge, a dining room, a ballroom or a concert hall, depending on the occasion and the arrangement of its furniture.

The steamboat's cuisine, included in the price of a cabin passenger's ticket, was an immensely important part of the cabin passenger's travel experience, not to mention the commercial success of the boat. Food and supplies were brought aboard at port cities and were also procured at landings along the river as the boat proceeded on its run. Chickens, pigeons, lambs and pigs were taken aboard as well as fruit, vegetables and fresh eggs. The animals were kept alive on the boat until the menu called for them to become dinner. Breads, pastries, cakes and other desserts were prepared in one of the boat's two galleys, the bakery ordinarily a part of the larboard (rivermen's usual term for the boat's left side) galley, and the meats and other courses prepared in the starboard galley. Meals were usually elaborate. One of the steamers offered its cabin passengers thirteen different desserts — six of them concoctions of custard, jelly and cream in tall glasses, and seven of them pies, puddings or ice cream. Another offered fifteen desserts. Some steamboats on the first day of their voyage served a dinner that was so heavy it left some passengers squeamish about taking in another full meal during the rest of the trip. The usual fare in early steamboat days was homey American food, but later, as the boats and their first-class passengers became more upscale, some French *haute cuisine* became *de rigueur*.

Less upscale was the boats' drinking water, which was served at every meal. It came, like the boat's water for its boilers and its passengers' washbowls, straight from the river, sediment and all. It was believed to be good for a person's system that way. Fortunately for the squeamish and finicky, coffee and tea, disguising the river water, were also served with meals.

At dinnertime a Mississippi River steamboat of standard elegance in the 1850s would provide as many as twenty-five waiters and attendants to take care of its passengers' prandial desires, and the saloon's lavish dining tables looked as if they were spread for an elaborate wedding reception. When it was time to take their places at the tables, the women passengers — the ladies — would process from their end of the saloon to music played by the boat's own band — whose members in some cases were all women — and when dinner was over, the ladies would march out to music. Everything considered, a cabin passenger's life aboard a Mississippi River steamboat was, as one old steamboat hand put it, "some powerful fine livin'."

The usual Mississippi River steamboat had four decks. The first, or low-

8 • The Sweet Life on the Mississippi

Stairway leading up from a typical steamboat's boiler deck, the deck above its main deck, to the hurricane deck, or promenade deck. At right in this photograph is the purser's, or clerk's, office. Passengers reached the boiler deck by climbing gracefully curving stairways that rose from the steamer's main deck near the bow of the boat (Library of Congress).

est, was called the main deck, which stood about four feet above the surface of the river. That was where the boat's machinery was mounted, with the boilers positioned forward and the engines positioned between the two huge paddle wheels. Some of the boats, of course, had a single paddle wheel, mounted on the stern. Also on the main deck were the galleys, space for freight and space for deck, or steerage, passengers, whose low fares entitled them to little more than passage and a sleeping spot on a cot, a bench or on the boards of the deck itself. Ten to eighteen feet above the main deck and reachable by a pair of curving stairways near the bow, was the boiler deck, or saloon deck, on which were the passenger staterooms, the barroom, the saloon — or main cabin — and the boat's offices. A promenade, like a porch, encircled the staterooms on the outside and could be accessed from the staterooms, from the saloon or from gangways, allowing cabin passengers to stroll or sit — on benches or chairs — and watch the passing scenery on the river and along the shore.

Interior of a stateroom. The grand saloon on either side was lined by staterooms with doors that opened into the saloon and also with doors that opened onto a porchlike promenade that encircled the boat's superstructure, allowing first-class passengers to stroll or sit and watch the passing scenery on the river and along the shore (Library of Congress).

Above the boiler deck was the hurricane deck, or promenade deck, with a cluster of cabins called the texas, in which the boat's officers were quartered. On top of the texas stood the pilothouse, or wheelhouse, some fifty feet above the water, presenting a commanding view of the river that lay before the bow of the boat. (A common belief is that the texas was so called because in the 1840s staterooms were named after the nation's states and the cabins that were occupied by the boat's officers, a recent addition to the steamboat's design, were named texas for the state of Texas, which in 1845 was the nation's newest addition. Another explanation for the name is that the officers' staterooms were the largest on the steamboats and therefore they received the name of the then-largest state, Texas. One other popular belief is that the cabins occupied by passengers and the boats' officers were called staterooms because they were named after states. However, the term stateroom, or state-room, had been used to mean a grand room in a palace or mansion before it was ever applied to a cabin on a steamboat or ship.)

8 • The Sweet Life on the Mississippi

A first-class ticket on a Mississippi River steamboat traveling upstream might cost as much as twelve and a half cents a mile, but about half that amount when the boat was headed downstream. In either case, fares were often negotiable with the captain, who in eagerness to take aboard as many paying passengers as possible would stop en route to pick them up when they hailed him from the shore, day or night.

For passengers who were unable to foot the expense of a first-class ticket, travel on a Mississippi River steamboat was considerably more austere and mean than it was for cabin passengers. Negro slaves and steerage-class white passengers, whose passage from New Orleans to St. Louis might cost as much as three dollars each, were quartered on the main deck, finding room wherever they could in between stacks of freight, forbidden to ascend to the upper decks. For sleeping, they brought their own bedding or did without. The boat provided a stove for cooking, but the deck passengers had to supply their own food, which was usually sausage, dried fish, or crackers and cheese — and a bottle of whiskey to wash down all that dry food. The deck passengers included farmers who had given up trying to wrest a living from poor soil in the eastern U.S. and were seeking more promising farmland in the West, and immigrants straight from Europe, seeking new lives in a new land, looking for a job and a place to start. Some others were simply restless individuals on the lookout for something better, something different, often bringing their wives and young children along with them on the quest. Others were peddlers, traveling with their wares from town to town, wherever the boat would stop.

Still others, particularly in the Mississippi steamboat's early days, were flatboat or raft crewmen who had come down the river on their vessels, which had been broken up and sold as lumber at their destination, and were returning to their homes upriver aboard a steamboat. They were a raffish bunch who treated the steamboat ride as a boisterous vacation, swilling rum and singing and shouting and firing their pistols into the air throughout much of the night. The negro slaves traveling as deck passengers, some of them bought in New Orleans and on their way to new locations, some being taken to New Orleans to be sold, were, like the flatboat crewmen, also temporarily free from their usual hard work, and many of them made the most of their trip, singing and dancing and generally celebrating the time aboard the boat as if it were a holiday. Charles Dickens, the British novelist, traveled aboard a Mississippi River steamboat in 1842 and complained that the passengers on the main deck kept him awake at night with their noise, shooting guns and singing hymns.

Music was one of the big attractions of Mississippi River steamboats. A brass band or an orchestra became standard equipment on the boats. It played for passengers in concerts and for dances during the voyage and it played for townspeople when the boat docked. Perhaps even more enjoyable to passengers was the music made by the free Negroes who worked as waiters, barbers, porters and deckhands aboard the steamboats. "They played stringed instruments," one observer commented, "and sang as only they could play and sing those haunting, joyfully sad melodies and hymns."[7] The calliope, or steam organ, or steam piano, was invented in 1855 (by Joshua C. Stoddard of Worcester, Massachusetts) and, although intended by its inventor to replace church bells, it soon found itself adopted by steamboat owners, who mounted the instruments on the exterior of the boats and had them played to charm not only the boats' passengers but people along the shore, the distinctive, cheery sound audible many miles away from the river.

The calliope, however, received mixed reviews once passengers had experienced it while aboard a steamboat. One long-time pilot on the upper Mississippi, George Byron Merrick, although acknowledging that music was important to passengers, did not think the calliope's tones made the sort of music that boosted the boats' passenger business:

> In the flush times on the river all sorts of inducements were offered passengers to board the several boats for the up-river voyage. First of all, perhaps, the speed of the boat was dwelt upon.... After speed came elegance — "fast and elegant steamer" — was a favorite phrase in the advertisement....
>
> After elegance came music, and this spoke for itself. The styles affected by river steamers ranged from a calliope on the roof to a stringed orchestra in the cabin [saloon]. The "Excelsior," Captain Ward, was the first to introduce the "steam piano" to a long-suffering passenger list. Plenty of people took passage on the "Excelsior" in order to hear the calliope perform; many of them, long before they reached St. Paul, wished they had not come aboard, particularly if they were light sleepers. The river men did not mind it much, as they were used to noises of all kinds, and when they "turned in" made a business of sleeping. It was different with most passengers, and a steam piano solo at three o'clock in the morning was a little too much music for the money. After its introduction on the "Excelsior," several other boats armed themselves with this persuader of custom; but as none of them ever caught the same passenger the second time, the machine went out of fashion. Other boats tried brass bands; but while these attracted some custom they were expensive, and came to be dropped as unprofitable.
>
> The cabin [saloon] orchestra was the cheapest and most enduring, as well as the most popular drawing card. A band of six or eight colored men who could play the violin, banjo, and guitar, and in addition sing well, was always a good investment.... They also played for dances in the cabin, and at landings sat on the guards and played to attract custom. It soon became advertised abroad which boats carried the best orchestras, and such lost nothing in the way of patronage.[8]

After-dinner activities aboard the steamers varied widely — and simultaneously. Passengers were allowed to amuse themselves as they pleased, so long as they did not infringe on the rights of others and did not interfere with

8 • *The Sweet Life on the Mississippi* 111

A steamboat calliope. Music was one of the big attractions of Mississippi River steamboats, and the calliope, invented in 1855, soon became a standard feature on steamers. It was mounted on the boats' exterior, and its distinctive, cheery sounds charmed not only the boat's passengers but people along the shore (Library of Congress).

the crew and the workings of the boat. "There might frequently be seen in the ladies' cabin a group of the godly praying and singing psalms," one traveler recalled, "while in the dining-saloon, from which the tables had been removed, another party were dancing merrily to the music of a fiddle, while farther along, in the social hall, might be heard the loud laughter of jolly carousers around the drinking bar, and occasionally chiming in with the sound of the revelry, the rattling of money and checks, and the sound of voices at the card-tables."[9]

Many steamboats had a rule that prohibited gambling after 10 P.M., but the rule was largely ignored, and it was not unusual for card games to last through the night and into the dawn of a new day. Some boats posted signs warning that gentlemen who played cards for money did so at their own risk. In the convivial atmosphere that prevailed after dinner in the saloon, members of the crew—"uncouth pilots, mates, and greasy engineers"[10]—sometimes joined well-dressed passengers at the card tables. The most common card games were poker, brag (similar to poker), whist, Boston (which required two decks of cards), and old sledge (also called seven-up). Other popular games included *vingt-et-un* (or blackjack), chuck (or chuck farthing, a coin-tossing game), three-card monte and faro.

Of all the passengers who ever boarded a Mississippi River steamboat, none were more remembered, or more written about, than the professional gamblers. At first they were regarded merely as very good players and accepted by fellow passengers as such. "The card tables of a steamer were free to all persons of gentlemanly habits and manners," George Byron Merrick wrote. "The gambler was not excluded from a seat there on account of his superior skill at play; or, at least, it was an exceedingly rare thing for one person to object to another on these grounds. Pride would not permit the humiliating confession."[11] Curiously, men who refused to associate with gamblers in ordinary circumstances ashore felt themselves in no way compromised by sharing a card table with them on a Mississippi River steamboat.

Pots were not big on the upper Mississippi, the playing passengers not being the wealthy planters that many passengers on the lower Mississippi were. Some did come aboard wearing broad money belts, though, laden with twenty-dollar gold pieces, and gamblers were usually satisfied to pick up two or three hundred dollars a week from those gold-bearing passengers.

Collusion was common. The professional gamblers often worked in pairs, coming aboard separately, pretending not to know each other, not speaking to each other until introduced, usually by an intended victim. They were convincing actors, playing a variety of roles to help lure suckers into a game. "At different times they represented all sorts and conditions of men—settlers, prospectors, Indian agents, merchants, lumbermen, and even lumber-jacks," Merrick wrote, "and they always dressed their part, and talked it, too. To do this required some education, keen powers of observation, and an all-around knowledge of men and things. They were gentlemanly at all times — courteous to men and chivalrous to women. While pretending to drink large quantities of very strong liquors, they did in fact make away with many pint measures of quite innocent river water, tinted with the mildest liquid distillation of burned peaches.... They kept their private bottles of colored water on tap in the bar, and with the uninitiated passed for heavy drinkers."[12]

The professionals apparently cooperated in a sort of a gamblers' fraternity that in effect granted informal franchises to particular individuals to work particular boats and discouraged encroachment by one gambler on the territory of another. Gamblers did occasionally switch from one steamboat to another, but only by agreement with their affected brethren.

Over time, the professionals developed a successful *modus operandi*. Once they had lured one or more victims into "a friendly game," the professionals, in the early hands of the game, would deliberately and cheerfully lose large pots to each other, and when the game had proceeded to the point where the intended victims felt comfortable and confident, one of the professionals would announce that the boat had reached his town and would disembark at, say, Prescott, Wisconsin, or Hastings, Minnesota, or Stillwater, Minnesota, and his partner would continue on to St. Paul, with the intended victims still at the card table with him, now ready to be fleeced.

"The chief reliance of the gamblers," steamboat traveler John Morris related, "lay in the marked cards with which they played. No pack of cards left the bar until it had passed through the hands of the gambler who patronized the particular boat that he 'worked.' The marking was called 'stripping.' This was done by placing the high cards — ace, king, queen, jack, and ten-spot — between two thin sheets of metal, the edges of which were very slightly concaved. Both edges of the cards were trimmed to these edges with a razor; the cards so 'stripped' were thus a shade narrower in the middle than those not operated upon; they were left full width at each end. The acutely sensitive fingers of the gamblers could distinguish between the marked and the unmarked cards, while the other players could detect nothing out of the way in them."[13] The professional gambler might spend hours stripping the cards in his stateroom, then replace them in their cartons, reseal the cartons and return them to the bar to be repurchased later, the bartender obviously being in collusion with the gamblers.

The deft hands of the professional gambler, who by the 1850s was perceived by Mississippi steamboat passengers as an *ipso facto* card sharp — rather than merely an unusually skillful player — were the key to his sure-thing success at cards. Dealing from the bottom of the deck in order to give predetermined hands to anyone at the table, was a technique believed to have been developed by a man named Wilson and first made its appearance on the Mississippi and other western rivers around 1834.

Detection, or even suspicion, of cheating was the chief occupational hazard of the Mississippi River steamboat gambler. Stories about threats to the gambler's life abound. One gambler, who had been particularly successful happened to overhear several of his victims conspiring to kill him and take back the money they had lost to him. He slipped away and found a place to hide near the pilot house, then bribed the boat's pilot to have him pull over close to the river bank at the first opportunity and let the gambler jump off the boat. The pilot did so, and the gambler leaped from the boat. He landed in shallow water and sank to his waist in the mud of the river bottom, trapped by the muck while his angry victims, having discovered his escape, began firing pistols at him. The steamboat continued on its way, and the gambler was soon out of range of the gunfire. He was finally rescued by slaves who had been working in a nearby field and were drawn to the river bank by the sounds of the gunfire. Responding to his yells for help, they got a long pole and pulled him to safety. He then waited on the bank for another passing steamboat to stop and pick him up.

That same gambler, George H. Devol, who had begun his career on the Mis-

George H. Devol, whose daring exploits made him one of the most famous Mississippi River steamboat gamblers. Many steamboats had a rule that prohibited gambling after 10 P.M., but the rule was largely ignored, and it was not unusual for card games to last through the night and into dawn of the next day. The most common card games were poker, brag, whist, Boston, blackjack and chuck (Library of Congress).

sissippi River as a steamboat cabin boy, was later in a similarly dangerous situation aboard another boat he was working. The men he had fleeced had become drunk and set out to find him and recoup their losses. He came out of his hiding place on the boat long enough to find some dirty clothing, which he put on, then smeared his face with grime and mixed in with the boat's roustabouts on the main deck. When the boat tied up at its next stop, he hefted a piece of freight and fell in with the roustabouts filing down the stage, hauling freight ashore. Thus he escaped once more, while his menacing victims searched for him on the upper decks.

In his memoir Devol told his own story of outwitting a desperate passenger who pulled a pistol on him after losing his entire bankroll to Devol playing monte, a game in which three cards are placed on the table face up and the bettor, after selecting one of the cards and having the dealer shuffle and manipulate them face down, then must pick from the three face-down cards the one he had selected:

> I was playing monte one night on the *Robert E. Lee* when a fellow stepped up to the table and bet me $800.... When he had lost his money and spent a few moments studying, he whipped out a Colt's navy [pistol] and said, "See here, friend, that is all the money I have got, and I am going to die right here but I will have it back."
> I coolly said, "Did you think I was going to keep the money?"
> He replied, "I knew very well you would not keep it. If you had, I would have filled you full of lead. I am from Texas, sir," and the man straightened himself up.
> Pulling out a roll of money, I said, "I want to whisper to you." He put his head down, and I said, "...I didn't want to give you the money before all these people because then they would all want their money back, too. But you offer to bet me again, and I will bet you $800 against your pistol."
> That pleased him. "All right," he said, and the $800 and the pistol went up in my partner's hands. Over went the wrong card. I grabbed the pistol, and told my partner to give me the stake money. Pulling the gun on him [the Texan], "Now," I said, "you've acted the wet dog about this and I will not give you a cent of your money, and if you cut any more capers, I will break your nose."[14]

The gambler wasn't always the winner, though. He could be outsmarted occasionally. The hero of one story, apparently an old one on the river, was a bank clerk who left New Orleans bound for Pittsburgh on bank business, carrying $100,000 in cash in his trunk. Several professional gamblers found out about the clerk's mission and bought tickets on the same boat with him. Once the trip started, it wasn't long before the gamblers had drawn the clerk into a game of brag, at which they allowed him to win several hands to set him up for the kill. At what they figured was the right moment, he was dealt a very good hand, and one of the professionals was dealt an even better hand.

The betting went back and forth between the clerk and the professional until at last the clerk had bet all the money that he had on the table. At that point the gambler raised him five thousand dollars and when the clerk said he was out of money and asked the gambler to show his hand, the gambler refused and demanded the clerk come up with five thousand dollars to see his bet or forfeit the pot.

"I go you five thousand better and give you five minutes to raise the money," the gambler told him.

The clerk slowly got up from the table and strode to his stateroom, went in and unlocked his trunk, then returned to the table with a package containing the money that he was taking to Pittsburgh. "You will not give me a sight for my money?" the clerk asked the gambler.

"No, sir," the gambler replied. "I went five thousand dollars better and gave you five minutes to raise the money. One minute of the time remains."

"Then, sir," the clerk declared, tossing the money package onto the table, "I see your five thousand and go you ninety-five thousand dollars better—and give you five minutes to raise the money!"

Unable to come up with that huge amount, the gambler and his partners abruptly withdrew from the table, leaving the pot—and the money package—to the clerk. At the boat's next stop, the gamblers got off to return to New Orleans, outmaneuvered and several thousand dollars lighter, but ready to take a new ride on another grand steamboat.

• 9 •

The Hard-Working Life

William Wells Brown was born into slavery in Lexington, Kentucky, in 1814 and when he was in his twenties, he worked aboard the Mississippi River steamboats *Missouri, Enterprize* and *Chester*, and on the Missouri River steamer *Otto* for a while. In 1847, after fleeing to freedom in Ohio and later having found a home in Boston, he wrote his autobiography, giving glimpses of life on the Mississippi as he saw it.

Brown was the son of a white plantation owner and a black slave woman and was sold as a child to a relative of his father. He moved with his new master from Kentucky to Missouri and was hired out to a Major Freeland, who in turn hired him out to work aboard the steamer *Missouri*, which ran between St. Louis and Galena, Illinois. In his book, a slender volume titled *The Narrative of William W. Brown, a Fugitive Slave*, Brown calls that steamboat assignment "the most pleasant time for me that I had ever experienced." He later was hired out to the captain of the *Enterprize* to work as a waiter. "My employment on board," he wrote, "was to wait on gentlemen, and the captain being a good man, the situation was a pleasant one to me — but in passing from place to place, and seeing new faces every day, and knowing they could go where they pleased, I soon became unhappy, and several times thought of leaving the boat at some landing place and trying to make my escape to Canada."

Although he apparently received little or no ill treatment himself while working as a waiter, he got a close look at how other slaves were treated and the conditions in which they were hopelessly trapped. It was the cargoes of fellow slaves that bothered him most. "On our downward passage," he recalled, "the boat took on board, at Hannibal, a drove of slaves, bound for the New Orleans market. They numbered from fifty to sixty, consisting of men and women from eighteen to forty years of age. A drove of slaves on a southern steamboat, bound for the cotton or sugar regions, is an occurrence so common that no one, not even the passengers, appear to notice it, though they clank their chains at every step.

"There was on the boat a large room on the lower deck," Brown wrote, "in which the slaves were kept, men and women, promiscuously—all chained two and two, and a strict watch kept that they did not get loose; for cases have occurred in which slaves have got off their chains, and made their escape at landing-places, while the boats were taking in wood—and with all our care, we lost one woman who had been taken from her husband and children and having no desire to live without them, in the agony of her soul jumped overboard and drowned herself. She was not chained." The part of the boat where the slaves were kept, Brown noted, giving an idea of the conditions under which the slaves traveled, was almost impossible to keep clean.

Brown was but one of many slaves who worked aboard Mississippi River steamboats. Historian Thomas C. Buchanan posits that if the crews of the 93 steamboats docked at St. Louis in September 1850 were representative of the 700 to 1,000 steamboat crews working on the western rivers by the middle 1800s, there were about 2,000 to 3,000 slaves and 1,000 to 1,500 free blacks at work on western steamboats at any given time. "The crew lists of these 93 boats," Buchanan reports, "indicate that 230 free blacks (6 percent) and 441 slave workers (12 percent) out of a total workforce of 3,627 toiled on these vessels."

The remainder of the steamers' crews were composed of American-born whites (43 percent), Irish-born (24 percent), German-born (11 percent) and 3 percent from an assortment of other foreign countries. The figures also show that 57 percent of steamboat crewmen worked as deckhands, 20 percent were cabin workers, and 1 percent were independent contractors, working for themselves as barbers or bartenders. The 2 percent of the crew members who were women worked as chambermaids.[1] The rest of the crew members were the boats' officers.

In the early days on the Mississippi practically all crewmen were white, but according to Herbert and Edward Quick, authors of an early twentieth-century book on Mississippi River steamboats, "Gradually, the negroes replaced all others as deck hands. They began as servants in kitchen and cabin and the more brawny found jobs as firemen.... As waiters their grins and native flattery were more pleasing to the officers and passengers than the grim condescending attendance of the whites; as cooks they were more satisfactory, and as firemen they would put up with more heat and abuse. They supplanted the white stewardess....

"It was not long until the happy, unworldly negroes made up more than half the crew of the steamboat. They cooked and served the meals, made up the bunks, stoked the fires and rolled the freight up and down the gangplank, working as stevedores in the hey-day of high wages for no more than fifty dollars a month. But they were free, and perhaps better off than the slaves who were always to be seen traveling on the main deck, having been bought by a slave dealer who was taking them to sell somewhere else."[2]

The racial and ethnic mixes of steamboat crews often made for explosive situations aboard the boats, as evidenced by the stories of racial violence sometimes published by newspapers. In June 1839 the *Picayune* in New Orleans reported that "a coloured man casually employed on board the *Maid of Orleans*, was wounded with a knife and much beaten on Friday night. The cause, we learn, was his attempting to eat supper with the white 'hands' on board."[3] In another case reported by the *Picayune,* this one resulting in the death of a black crewman, "One of the white deck hands undertook to beat the negro and ... another one drew a knife and stabbed the negro."[4] The reason for that altercation was not reported. Often the cause of fights was little more than racial bigotry. A free black who was a cook aboard the steamer *Aunt Letty* in 1857, drawn into a fight between white deckhands and a black fireman, bellowed a menacing challenge as he went to the fireman's aid: "I can whip any goddamn white-livered son of a bitch on the boat if they would give me a white man's chance!" Minutes later, the white deckhands having been reinforced by the boat's mate and at least one other officer, a free-for-all erupted. White crewmen attacked a group of blacks that included the cook, now armed with a knife, the steward, also armed with a knife, and several waiters. After the initial clash, the black crewmen, except for the steward, fled the fray. The steward, holding a sixteen-inch knife from the kitchen, stood alone to face the white crewmen, who quickly overwhelmed him and were about to cut his throat when the boat's clerk, armed with a gun, threatened to shoot the first man that moved toward the steward. That ended the confrontation, at least temporarily, but the threat of violence remained. Not long after that incident, six of the *Aunt Letty*'s free-black crew members quit their jobs and went ashore after the boat's mate declared that he would "kill every nigger on board the boat."

Much of the racial animosity between crew members arose from white resentment over steamboat owners' employment of slaves and free blacks at a cost cheaper than the wages of white crewmen, which inspired in white workers a fear that their jobs would be lost to blacks. In an effort to reduce racial friction and violence aboard their vessels, some captains resorted to hiring cabin crews that were either all white or all black.

One usual device for avoiding friction between black and white crewmen, in accordance with the social norms of the time, was to segregate crew

Cotton bales piled high on the decks of the steamer *William Garig*, docked at Baton Rouge. The bales were brought aboard the steamboats by roustabouts and were stacked by the boat's deckhands under the direction of the boat's first mate. Deckhands also stowed other freight, either in the boat's hold or on deck, and retrieved it when the freight's destination was reached (Library of Congress).

members during meals, not only by race but by the crewmen's status in the steamboats' work force. On the lower Mississippi, cabin passengers generally were served first, then, at separate sittings, the boats' officers were served, then white servants and black passengers, then black crew members. Black passengers were not allowed to occupy staterooms and were restricted to the boats' main deck. One contemporary account reports that the dining sequence was as follows: "the passengers, one set after another, and then the pilots, clerks, mates and engineers, and then the free colored people, and then the waiters, chambermaids, and passengers' body servants."[5]

Crewmen not included in any of those groups ate as best they could from the leftovers of diners who had had their fill. One account reports that the feeding procedure was for the roustabouts, the lowest workers in the steamboat crews' pecking order, to grab what they could from leftovers placed on the main deck in pans, meats in one, bread and cake in another and jellies and custards in a third. The low-status crewmen were summoned to their food with a cry of "Grub pile!" and after they had taken what they could, they would sit on the deck and eat.

Brawn was the main requisite for the roustabouts' job, which required the roustabouts to strong-arm freight onto and off of the boats. Nimbleness was a big plus, because of the danger of slipping off narrow gangplanks while carrying freight aboard and ending up in the river, which sometimes resulted in roustabouts drowning. Once brought aboard by the roustabouts, the freight, including weighty, cumbersome bales of cotton, was stowed by the deckhands, either in the hold of the vessel or on deck, under the direction of the mate. Stacking cotton bales was an important part of the deckhands' job, and until the 1840s, when the steam engine was adapted for hoisting the bales, it was done manually, using capstans to lift and precisely stack the bales. Deckhands also retrieved items of freight from stowage when they reached their destinations.

Deckhands were used, too, by the pilot to take soundings of the water's depth and to call back to him the measurements. The depth was measured in fathoms, six feet to a fathom, using a thirty-foot rope that was dropped overboard and allowed to sink to the river bottom. One detailed account of the sounding procedure describes the rope as weighted with a twelve-inch-long pipe about an inch and a half in diameter, into which a length of chain was inserted and held in place with lead that was poured into the pipe. The end of the rope was tied to a link of the chain that extended from the pipe. Markers were affixed to the rope, called a "lead line," to indicate the water's depth as the weight sank into the river. Four feet from the end of the line was a piece of white flannel that was woven into the rope. At six feet from the end there was a piece of leather attached to the line, and at nine feet was a piece of red cloth. At twelve feet — mark twain — there was a piece of leather split into two thongs. At eighteen feet — mark three — there was a piece of leather split into three thongs, and at mark four there was a leather strip with a round hole in it. Those markers were readily identified even in darkness by the leadsman, the deckhand manning the lead line, and he called out the depth as the weight sank and the line slipped through his hands. When the leadsman called, "Mark twain!" he was telling the pilot that the water was two fathoms, or twelve feet, deep. "Quarter less three" meant the water was a quarter of a fathom less than three fathoms, or sixteen and a half feet, deep. "Half twain" meant the water was two and a half fathoms, or fifteen feet, deep. A depth shallower than a "quarter less twain" was measured in feet. A depth of more than "mark four"—twenty-four feet—was reported by the leadsman as "No bottom!" That was the call the pilot yearned to hear.

Measurements differed according to whether the weighted pipe lay horizontal on the river bottom or stood upright, with only its end touching the bottom. In some cases that difference could be critical. When the pipe rested horizontal on the bottom, the measurement was known as "laying lead." When the pipe was vertical, it was called "standing lead."

When a depth measurement was desired by the pilot, he would sound a whistle or bell, once for the measurement to be taken on the starboard side, twice for the larboard side. The soundings would be taken about every one hundred feet as the boat moved through the water. The leadsman, who ordinarily was some distance from the pilot standing atop the boat in the pilothouse, had to make certain he could be heard, even through the noise of wind and rain. To do so, leadsmen used chants when they called out the measurements, singing their own tunes and rhythms that they felt expressed the various depths of the river, always holding their notes to be sure their reports of the measurements were understood. On the Mississippi's largest steamers the distance between the leadsman, down near the surface of the river, and the pilothouse was so great that the measurements were relayed from the leadsman to a deckhand posted closer to the pilothouse, who would then repeat the leadsman's call. When the steamer was out of danger, having moved into deeper water, the pilot with a signal would end the soundings.

For times when the river was unusually low there was a more cautious procedure for sounding, which is detailed by Samuel Clemens in *Life on the Mississippi*. The boat would tie up to shore, he explains, and the pilot who was not on duty would take a hand-picked crew of deckhands and with them would leave the steamer, board its yawl—or a special sounding boat if the steamer was big and posh enough—and row out ahead of the steamer, searching for a channel that would allow the steamboat to get over the menacing sand bar that lay before them. At night, the yawl crewmen would find their way through the river's blackness in the light of the yawl's lantern. In the pilothouse the pilot on duty would watch the sounding crew's movement through a telescope and by means of signals with the boat's whistle would direct the sounding crew to test the water's depth in one direction or another. When the yawl approached the shallow spot—the shoal place—the rowers would slow the yawl and the off-duty pilot would repeatedly measure the water's depth with a pole ten or twelve feet long until he found the shallowest place. He would then order the yawl stopped and have the crew drop a buoy overboard.

At that, the yawl's crewmen would stand their oars straight up to signal the pilot in the pilothouse, and he would give a blast of the boat's whistle to acknowledge the yawl's signal. "The steamer comes creeping carefully down," Clemens relates, "is pointed straight for the buoy, husbands her power for the coming struggle, and presently, at the critical moment, turns on all her steam and goes grinding and wallowing over the buoy and sand and gains the deep water beyond."[6] At night a paper lantern with a lighted candle in it was placed atop the buoy, its glow visible for a mile or more.

Or maybe the steamer doesn't make it across the sandbar, Clemens conceded. In that case the steamer would rest there on the sandbar until its crew sparred it off, a procedure that tediously lifted the boat, with the aid of steam power, and shoved it ahead inches at a time, until the boat finally escaped the sandbar after many hours or even days.

Sometimes a buoy was not used, and instead of heading for the buoy, the steamboat would simply follow the yawl at a safe distance and at a cautious speed. Clemens claimed there was often a great deal of fun and excitement that went along with sounding in the yawl, particularly if done on a beautiful summer day. But in the winter, he said, the cold and the danger took most of the fun out of it.

Other responsibilities of the deckhands included manning the ropes to tie up and to cast off at landing sites, as well as keeping the main deck clean and pumping out water that had seeped into the hold. Individual deckhands also served as night watchmen, taking turns at patrolling the deck during the night, alert for fire or any other danger to the vessel. The typical Mississippi River steamboat would have forty or so deckhands in its crew.

Firemen, the crewmen who kept the furnaces fueled, were part of the deck crew. Most were negroes. The work was so physically demanding that many boat officers considered the fireman's job to be for blacks only. One eyewitness described them at their labors. "The immense engine fires," she wrote, "are all on deck, eight or nine apertures all in a row ... like yawning fiery throats, and beside each throat stood a negro naked to his middle, who flung in firewood. Pieces of wood were passed onward to these feeders by other negroes, who, standing on a lofty stack of firewood, threw down with vigorous arms food for the monsters on deck."[7]

It wasn't only the stoking of the fires, moving four-foot-long logs from stacks on the deck and tossing them into the roaring furnaces, that made the job laborious; it was the hauling of wood onto the boats from riverside woodyards. Steamers burned six or more cords of wood a day, and the boats had to stop periodically to take on more fuel, often loading twenty or more cords onto the vessel each time it stopped for refueling. It was the firemen's task to hand-carry the wood aboard, sometimes aided by deck passengers who had paid discounted fares on the condition that they would help with wooding, as the wood-toting task was called. The firemen and whatever helpers were assisting them would line up in a long queue and file off the boat on a gangplank, or landing stage, trudge up the levee and gather into their arms several of the four-foot logs, turn around and plod back down onto the boat and stack their burdens in piles near the furnaces, then go back, again and again, for more armloads until the wooding was completed. That work, of necessity, was carried on day or night, wintry or warm, dry or raining. Firemen were also required to clean out the furnaces and haul away the accumulated ashes. For their demanding work they received wages slightly higher than those paid other members of the deck crew. Around the middle of the nineteenth century deckhands and roustabouts were paid about thirty dollars a month, and firemen were paid about thirty-five dollars a month.

Over time, a faster system of taking on wood aboard the steamers was developed. It involved the use of a flatboat, or scow, loaded with twenty or more cords of wood, to be towed alongside the steamer and the wood transferred from the flatboat to the steamer while the two vessels moved upstream together. A contract for the wood would be made in advance of the fueling, the date and time of the steamer's arrival being specified, and the flatboat would be waiting, day or night, for the steamboat, which would pause only long enough for the flatboat to be tied up to it. As the two vessels proceeded upriver, the steamer's team of wood bearers would toss or carry the pieces of wood from the flatboat to the steamboat. The whole process, employing as many as forty men, would take about thirty minutes, the steamboat continuing its movement all the while, and when all the wood had been transferred to the steamboat, the towlines were cast off and the flatboat, manned by two crewmen of its own, would be allowed to drift downriver back to the wood yard with the current, the two crewmen guiding it with a long steering oar at each end of the craft. The distance it covered would usually be no more than five or six miles. Steamers paid a premium for the wood when such wood-boats were used. The system was seldom employed on downriver trips, for short of being towed, there was no way for the flatboats to return to their woodyards.

Cooks were another group of crewmen who worked on the main deck—and not under the best of conditions. George Byron Merrick in his recounting of steamboat operations on the upper Mississippi reports that there was a popular saying that if you wanted to save on the meals that a passenger was entitled to on his trip, you would take him through the kitchen the first thing after he came aboard. The implication being, Merrick pointed out, that after seeing the food in the process of preparation, passengers would have little appetite for it when it came to their tables. Merrick claimed it would be unfair to the average steamboat steward to accept the saying as wholly justified, but, he acknowledged, "it would be stretching the truth to assert that it was without foundation." Steamboat kitchens in which meats and vegetables were prepared were cramped — usually measuring no more than ten by twenty feet — congested, messy and almost continually busy.

The cooks were required to turn out three meals a day for three to four hundred people, including passengers and crew, starting with an early-morning breakfast for which preparations began at 3 A.M. As soon as breakfast was over, the menu for dinner was written down by the steward and turned over to the chief cook, and efforts were begun to round up whatever was needed, often requiring time to take on provisions at the steamer's next stop. Pigs, lambs and chickens were taken aboard live and slaughtered and dressed on the boat by assistant cooks. Chickens, taken from their coops, their necks wrung or broken on the iron rim of a barrel, were plunged into a barrelful of scalding water drawn from the steamer's boilers and then had their feathers plucked from their steaming bodies before being eviscerated and their parts prepared for cooking, the process sometimes reaching a rate of a hundred and fifty chickens an hour, all done within the purview of the cooks and their bustling kitchens, reeking of many odors and aromas simultaneously.

On the other side of the boat was the bakers' and pastry cooks' kitchen, no bigger but far more inviting. It was, according to Merrick, the showplace of the boat. "Most stewards," he wrote, perhaps revealing then popular tastes in food, "are shrewd enough to employ pastry cooks who are masters of their profession, men who take pride not only in the excellence of their bread, biscuit, and pie crust, but also in the spotlessness of their workshops. They are proud to receive visits from the lady passengers, who can appreciate not only the out-put but the appearance of the galley. It is a good advertisement for the boat, and the steward himself encourages such visits, while discouraging like calls at the opposite side."[8]

Cooks sometimes worked eighteen-hour days, often in torturous heat, from early morning till the middle of the evening when all their pots and pans were again clean and put away for the next day's use. The cook designated as first cook was the head of the kitchen, directing and overseeing a staff that might total as many as four cooks, depending on the size of the boat and its passenger-carrying capacity. On small steamers there might be as few as two cooks, including the first cook, and those two did everything the kitchen operation required. The first cook also did the hiring and firing of his subordinate cooks and paid them their wages from funds provided by the captain. The man lowest in the cooks' pecking order was responsible for the most onerous kitchen duties — lighting the fires, washing pots and pans, and assembling the piles of leftovers from which ordinary crewmen ate. First cooks were able to bargain with captains for their pay and in the 1850s made as much as fifty dollars a month, sometimes more. Subordinate cooks received considerably less.

Above the main deck toiled the cabin crew, an assortment of workers whose primary duty was to make the cabin passengers comfortable and keep them happy. As important as any members of the cabin crew were the chambermaids, usually the only female members of the steamboat's complement of workers and usually numbering no more than two aboard an average size steamer. Not only did they keep the passengers' staterooms clean and orderly — which included making up beds, emptying washbowls and refilling water pitchers — they also washed the bed linen, towels, tablecloths and napkins, and often the passengers' and boat officers' clothes. The washing was done in wooden tubs, and the items were hung up to dry along the rails or on the main deck. Whatever needed pressing the chambermaids ironed with irons heated on coal fires and when all items were clean and pressed, the chambermaids returned the clothes to the passengers and the boat's linen to the steward. Working from early morning till late evening, chambermaids had good reason to believe the maxim that "woman's work is never done."

Their pay might run as high as twenty-five dollars a month, but they also received tips. Passengers tipped them for washing and ironing their clothes, for assisting women passengers with getting dressed and for other help. Slave workers ordinarily were not tipped, but passengers often couldn't tell who was and who wasn't a slave and tipped them regardless. One chambermaid, a free black woman, reported that she made six to seven dollars in tips on each trip between New Orleans and Natchez.

Like other members of the crew, chambermaids received what was in effect free room and board while working on the steamers. They slept on the floor of the main cabin, or saloon, after the cabin passengers had turned in for the night or, on trips when there were empty staterooms, they were allowed to sleep in the passengers' berths.

Sexual abuse was an occupational hazard for chambermaids. They were vulnerable to unwanted advances and even assault from ordinary crewmen, passengers and the boats' officers, including the captain. Slave women were particularly vulnerable, with little or no legal protection, but after the Civil War and the elimination of slavery, black women resorted to the courts to protect them and punish their abusers, not often with satisfactory results, however.

The boats' porters carried the cabin passengers' baggage aboard and off the boats, under the supervision of the clerk. They issued baggage-claim checks to passengers and stowed the bags that were not kept in the passengers' staterooms. While the boats were in motion, the porters had little to do, but at every stop where cabin passengers came aboard or disembarked, they leaped into action to assist.

Waiters were almost a separate class of crew members. Their jobs, more so than the others, put them in continued contact with cabin passengers, which included the traveling elite of the times, and thus required of the waiters a certain poise, intelligence and appearance, as was the case with William Wells Brown. Appropriately dressed to serve the boats' first-class passengers, waiters usually wore suits or dark pants and white coats. They set the dining tables, took food orders, served the food and drinks and bussed the tables. They also tended the coal stoves that warmed the saloon, set up cots in the saloon for cabin passengers who bought their tickets after the staterooms had already filled, sometimes helped the cooks with food preparation, ran errands for the steward and kept the saloon clean. They worked under the direction of the steward, who hired them. The job of steamboat waiter, along with that of steward and chief cook, carried prestige in the black communities along the river, and it often led to promotion, by the captain, to steward, a position highly prized and well rewarded.

Steamboat barbers were usually free blacks, though sometimes slaves, who contracted with the boat owner and rented space for their shops and worked for fees and tips. Bartenders, or barkeepers, were usually white, bright, young and personable. " It was required by their employers," George Merrick wrote, "that they be pleasant and agreeable fellows, well dressed, and well mannered. They must know how to concoct a few of the more commonplace fancy drinks affected by the small number of travellers who wished such beverage — whiskey cocktails for the Eastern trade, and mint juleps for the Southern."[9] Western men, Merrick claimed, took their whiskey straight, "four fingers deep."

Merrick also claimed that "in the old days on the river"— before the Civil War — whiskey was considered a necessity, including aboard Mississippi River steamboats. "It was a saying on the river," he wrote, "that if a man owned a bar on a popular packet, it was better than possessing a gold mine.... Men who owned life leases of steamboat bars willed the same to their sons, as their richest legacies."[10] It was not unusual for a person to hold leases on bars aboard several steamboats, hiring bartenders to operate them for him while he strictly supervised them. In the 1860s, Claiborne Greene Wolff, who had once been a steamboat steward, was the co-owner of the leases to bars on some thirty steamers.

Tending to confirm Merrick's assertion of the popularity of whiskey on Mississippi River steamers was the practice of providing a free supply of it to crewmen as they worked. Like rum on sailing vessels, whiskey rations for crewmen on Mississippi River vessels, first keelboats and then steamers, were a means of attracting workers to burdensome jobs as well as a way to ease the pains of their tasks. "We always gave our men, black or white, as much as they wanted," the captain of the *Ben Sherrod* admitted. "[We] kept a barrel of whiskey tapped on the boiler deck for them, have always done so, and let one of the watch draw for his mates. I have done the same for the last ten years."[11]

After the *Ben Sherrod* caught fire during a race between New Orleans and Natchez in May 1837, that same captain, named Castleman, denied that whiskey was part of the cause. "My acquaintances will vouch for my discipline about drunkenness being severe," he said. "Indeed I am generally blamed for being rigid with my hands." He conceded the truth of reports that his firemen were singing and dancing when the fire broke out, but stated that "they always do when on duty," apparently as a means of removing some of the hardness from the work.

• 10 •
Owners and Officers

On the Hudson and other rivers in the eastern United States steamboats were generally owned by companies, which paid captains to run them. Steamboats on the Mississippi and its tributaries, however, were in many cases owned by the captains who operated them. Alex Scott was one such Mississippi River owner-captain. He ordered a boat built and when it was put into service, running between St. Louis, Pittsburgh and New Orleans, he became Captain Scott and put himself to work at whatever tasks needed doing. When the boat was under way, he could usually be found on deck somewhere, often near the furnace doors, assisting the firemen. When the boat landed, he was up on the forecastle, assisting in one way or another the deckhands handling the freight. Ever alert to whatever was happening aboard his boat, he had a reputation for seldom sleeping while the boat was in operation.

On one rare occasion, though, he did fall asleep and became the victim of his crewmen's practical joke. A mild-mannered, good-humored man (whose harshest expression was "by the Lord Harry!"), Scott was well liked by his crewmen, but they sometimes took advantage of his gentle nature. One night as his boat, the *Majestic*, was steaming up the Mississippi, Scott took a seat on the capstan, one of his favorite spots, and while sitting there, dozed off. Several crewmen noticed him sleeping and gingerly turned the capstan halfway around so that instead of facing the jackstaff on the bow, Scott was facing the boilers while he slept. On a signal, the firemen opened all the fire doors at once, revealing the flames in the furnace, glowing brightly in the darkness, and at the same time they roused the captain. Seeing the light of the flames in front of him, he instantly concluded that another steamer was bearing down on his bow. He leaped from his seat on the capstan and yelled up to the pilothouse, "Stop her, Mister Pilot, or by the Lord Harry she will be into us!" When his crewmen broke into laughter, he realized his mistake and began laughing himself, enjoying the joke almost as much as the crew.

Some owners, such as James Dozier, from Nash County, North Carolina, were businessmen who were engaged in other enterprises and seeing opportunity in the steamboat business, entered it as well. As a young man Dozier moved to Paris, Tennessee, and began farming, became prosperous at it, then went into the mercantile business and later operated a tanning business with his father-in-law. After that he became a steamboat owner. By 1844 he had owned six steamers and with his sons he later became owner of at least three others. After the Civil War he moved to St. Louis, where he founded a large bakery.

Another steamboat entrepreneur was St. Clair Thomasson of Louisiana, who was a partner in a wholesale dry goods business and in 1843, with his business partner, Theo Shute, built the steamer *Baton Rouge*, which he captained and operated between New Orleans and Vicksburg.

Joseph Throckmorton was an owner-captain whose business career went the opposite way. He first owned several steamers that he operated on the upper Mississippi in the 1830s and 1840s, but around 1850 he decided to leave the river and try something else. He went into the insurance business in St. Louis and when that venture failed, he returned to what he knew best, steamboating. After he died in 1872 at age 72, a biographer poetically informed his readers that Throckmorton had "crossed the river that ferries but one way."

In some cases owner-captains started their careers as boat-builders and knew steamboats from the inside out. James Ward first worked in a boatyard and from that job went to work as a carpenter on the steamer *Ione* and then on the *Amaranth*, operating out of St. Louis. In 1844 with three partners he built the steamer *St. Croix* and joined its crew as mate. He then sold his interest in the *St. Croix* and with two partners built the *St. Peters*, which he served as captain, running between St. Louis and Dubuque.

Many owner-captains in the 1830s and 1840s had first operated keelboats on the Mississippi and had then graduated to steamboats when they saw the steamers' greater potential and superior working and living conditions. W.J. Koontz, born in Columbiana County, Ohio, in 1817, began his career on a keelboat that was owned by his brother. He later became a steamboat pilot, then a captain and then an owner-captain. He volunteered his services to U.S. general George B. McClellan during the Civil War, was made a commodore, posted to St. Louis and placed in charge of river transportation for Union troops.

L.T. Belt, born in St. Clair County, Illinois, in 1825, one of twelve children, was another owner-captain who advanced from keelboats to steamers. He and his brother Francis gave up keelboating and bought the steamer *Planter* in 1847 and operated it on the Mississippi between New Orleans and St. Louis. From that beginning Captain Belt eventually became president of the New Orleans and Bayou Teche Packet Company, operating steamers on Bayou Teche in Louisiana. Belt had other laudable accomplishments. For years he was superintendent of the Sunday school at the Rayne Memorial Methodist-Episcopal Church in New Orleans and was also, his biographer pointed out, an unusually dutiful son: "For the twenty years previous to her death, he never, but once, no matter how great the distance, failed to visit his aged mother on her birthday."

Many other Mississippi River captains, of course, were not boat owners, but rather salaried employees of the companies that owned the steamers of which they were masters. Charles S. Rogers, born in New Hampshire in 1816, moved to St. Louis when he was 22 and got a job as clerk on a Mississippi River steamboat in 1842. From that job he rose over time to captain and later became president of Naples Packet Company, which operated twenty-three steamboats and a number of barges and wharfboats. Henry W. Smith began his working life at a country store in Missouri in 1855, then took a job as second clerk on the *General Lane*, a Missouri River steamer. He advanced to captain of another Missouri River steamer and later became captain of a steamboat running between New Orleans and St. Louis. After the Civil War he became president of the St. Louis & Memphis Packet Company and distinguished himself for building the company's steamboat fleet into one of the fastest and finest on the Mississippi.

One of the more unusual captains was William F. Davidson, who late in life became a Christian, got caught up in the temperance movement and thereafter banned bars on the several steamers he controlled, which operated between St. Paul and St. Louis. William Dean, whose career spanned thirty years as a pilot and captain, was so conscientious in his Christian beliefs that he would not operate his boat on a Sunday. Another unusual captain was Mortimer Kennett, who was something of a violin virtuoso and every evening played his violin for his passengers. "No navigation," it was said of him, "was too difficult or night too dark to induce him to decline the very pleasant duty of entertaining his passengers with the sweet strains of his violin."[1]

The tasks of the owner-captain were varied and many. He was constantly alert for anything that needed his attention. "He watched the work of his engineer in the care of the machinery and the purchase of fuel, oil and other supplies. He checked up [on] his steward in the matters of food, tableware, linen, and all the other things necessary to the operation of a passenger boat. Under his eye the mate attended to the loading, stowage, and unloading of freight. And with the clerk he went over the boat's accounts, decided where he was making money and where he was losing, and regulated his policies accordingly."[2] He also set freight and passenger rates, processed claims and made out the boat's schedules, along with taking care of a host of other things that needed doing, including receiving complaints from his crew.

David Hiner was a captain who, according to an old story told about him, had his own way of dealing with his crew's complaints. Not long after taking command of a fine steamer, he was practically besieged by members of his crew asking for more or better than what they had. The mate wanted a new hawser. The steward wanted a new cooking stove. The engineer wanted a new doctor engine. The porter wanted a new badge for his hat. The chambermaid wanted a glass pane installed in the window of her stateroom. Fed up with the crew's gripes, Captain Hiner decided that what he wanted was a new crew. He fired the old one and hired a whole new one.

The captain's job required that he exhibit the sort of pleasing, outgoing personality that would let him easily meet and mingle with shippers and passengers. It was his ability to attract and keep customers — as well as his business ability — that determined the financial success of his boat. To the cabin passengers aboard his vessel he was a gracious host. In many cases he developed lasting relationships with passengers and shippers. He often got to know not only his shippers but their families. His customers in turn often developed a friendly interest in him and his boat and once becoming comfortable with both, they seldom took their business elsewhere.

One big reason for shifting their business to another boat was a rate cut by a rival steamer, an act usually aimed at putting a competitor out of business. Owners faced with such competition often abandoned their boat's ordinary run and shifted their operations to another section of the river or to another river entirely rather than engage in a ruinous rate war. Others, though, accepted the challenge of the rival steamer and fought back, cutting their rates to match or beat the rival's. Rate wars often continued until one of the boats went broke and had to be sold to satisfy the debts resulting from its losses. The owner-captain, however, was usually soon back in business with another boat.

To members of the crew the captain was "captain" or "sir" when they spoke to him, but they were more likely to refer to him as "the old man," no matter his age, when speaking *about* him, out of his hearing. "The old man" was a term of respect, even endearment, earned through the captain's deeds and words. Over the members of his crew he held complete authority. "After leaving port, the captain on the river was as autocratic as his compeer on the ocean," according to veteran steamboat pilot George Byron Merrick. "He might without notice discharge and order ashore any officer or man on board, and he could fill vacancies en route to any extent ... subject to the approval of the owner or manager on arrival at the home port."[3]

There were boundaries, however, which the captain, even the owner-captain, was loath to cross. Steamboat custom, tantamount to law, was that captains did not ever interfere with their pilots in the operation of the boat, even if the captain was himself a licensed pilot and was certain he knew better in certain situations than the man in the pilothouse. Neither was it a good idea to meddle with the mate during the performance of his duties or the engineer or the chief clerk during the performance of theirs. An emergency that threatened the safety of the boat was about the only good reason for the captain to interfere with those officers while they worked, and any captain who interfered without good reason suffered the loss of respect not only aboard his own boat but on others that plied the Mississippi. The captain might—and did—though, call an erring officer into his office and in private, away from the ears of crewmen and passengers, order him never to repeat whatever transgression had aroused the captain's concern.

A visit to the captain's quarters was not always an unpleasant experience. The captain's cabin, or suite, was ordinarily the most spacious and best furnished of all the steamboat's accommodations. It was usually at the forward end of the texas deck and it served as an office, sitting room, meeting room and private dining room; connected to that multi-purpose space was the captain's bedroom. The captain used the multi-purpose space at times for conferences with his officers, for entertaining VIPs, or for merely dining, alone or with favored passengers. For passengers, an invitation to the captain's quarters, to dine or be otherwise entertained, was a high honor and a special treat. The view there was spectacular. Large windows on the front and two sides of the multi-purpose room gave a panoramic vista of the river and the shore on both sides as the boat steamed ahead.

While in his cabin the captain could monitor his pilot in the pilothouse above him, detecting his movements at the wheel and hearing him pull the bells that signaled the engine room as the boat moved into dangerous water, often giving cause for the captain to return to the deck. During times of emergency, the captain was expected to keep a cool head. He, with the pilot, had to quickly decide where to beach the boat when fire broke out aboard the vessel. Moments of indecision or inaction could mean disaster for the boat and its passengers and crew. "In case of snagging, or being cut down by ice," Merrick states, "it is his first duty to save his boat, if possible, by stopping the break [in the hull], at the same time providing for the safety of his passengers by beaching her on the nearest sand-bar."[4] The code of conduct that required the master of an ocean-going vessel to be the last man to leave it when it was burning or sinking applied equally to the steamboat captain, and many a steamer captain, according to Merrick, lost his life adhering to that code.

Among the owner-captain's prerogatives was the naming of the vessel. The names chosen were often those of individuals, some prominent, such as *Robert E. Lee, Tecumseh, Henry Clay, Robert Fulton* and *John Adams*, and some not, such as *A. T. Lacy, Henry J. King, Hiram Powers, Fannie Lewis* and *Grace Darling*. Place names were also popular, such as *Wyoming, Honduras, Pittsburgh, City of Alton, Denmark* and *Hawkeye State*. Outside those two categories, almost any name was possible — such as *Iron Duke, Resolute, Little Giant, Rover, Vixen, Editor, Post Boy, Fairy Queen, Fire Canoe, War Eagle, Harmonia, Greek Slave, Northern Light, Time and Tide* and *Ocean Wave*. There was one group of names, though, that some owner-captains avoided, with what they believed to be good reason. Captain John N. Bofinger, whose Mississippi River experiences extended over forty years, explained the abhorrence of those names:

> I do assert that, with barely an exception, that all steamboats built and run on the Mississippi River and its tributaries, whose names commenced with the letter M, were either burnt, sunk, exploded or unsuccessful as an investment to their owners. You can look over a long list of *Missouri, Mississippi, Mary, Michigan, Marie, Monarch, Mediator*, etc., and you will find that they met the fate of one as above indicated.
>
> Over thirty years ago, Capt. John Pierce built the *Metamora*. I tried my best to persuade the captain to name his boat some other name, and gave him my reasons, going over a large number of boats whose name had commenced with the letter M. He laughed at what he called a superstitious notion of mine and called his boat the *Metamora*. She was a great success, but sank above Choctaw island while she was in her prime. Capt. Charley Davis, about the same time, built a splendid Cincinnati and New Orleans boat. Davis, like his old partner, Pierce, would not listen to my idea, launched and christened her the *Midas*. She sank in the bend above Island 16. Capt. Joe Brown built the *Mayflower* sometime during the fifties. Long before she was launched I tried to talk him out of calling the boat by that name — no use. She was burned at Memphis.
>
> Our old townsman, Norman Cutter, Esq., bought a hull that had been built at Hannibal. Her cabin and machinery was put on at St. Louis, where she was finished.... It was the owner's intention that I should have taken charge of the *Charles Belcher*, which was the name Mr. Cutter gave her about a month before the *Belcher* was ready to start on her first trip. I accidentally found out ... that she had been launched and christened *Magnolia*. That was enough for me. Nothing could have induced me to have taken charge of the *Belcher*. She was burned on her sixth trip at New Orleans.
>
> I could name hundreds of instances to show the fatality that seems to shroud the steamboats whose name commenced with the letter M, but will content myself with giving one more instance. I was in New Orleans in May, 1875, where I met Capt. Frank Hicks and his clerk, Mr. Alf. Grissom, who were at that time building a hull at Metropolis, Ill. They talked of calling her the *Mary Bell*. I did my level best to persuade them not to call her that name or any name

that commenced with M; gave them my reason and recited many instances of losses, etc, all to no good; the boat was called *Mary Bell*, made but a few trips and burnt with a full load at Vicksburg.

I do not pretend to give any reason why a steamboat's name commencing with the letter M should be any more unlucky than one commencing with any other letter, but the fact still remains, superstitious or not.[5]

There were, of course, M-named steamers that escaped the supposed M curse and lasted the normal life span of a Mississippi River steamer. Some that did so were the *Majestic, Mary Hunt, Music, Mary Foley* and, perhaps the most outstanding exception to the rule, the *Mollie Mohler*, the double M of its name proving no bane as it operated successfully for many years, and after its superstructure was dismantled, a new boat was built on its hull.

Another superstition among steamboatmen was that a name with six letters in it was also unlucky.

The job of Mississippi River steamboat captain was not for men only. Women could take the rigorous examination administered by the U.S. Department of Commerce Steamboat Inspection Service, the same test that was required of men, and if they passed, they became qualified and licensed to operate steamboats on the river. Blanche Douglass Leathers, wife of Bowling Leathers, the son of *Natchez* owner-captain Thomas P. Leathers, was one of the most colorful and most written about women to captain a Mississippi River steamboat. Blanche married Bowling in 1880, moved into the captain's cabin with him — rather than wait on land for him to return from his steamboat runs — and by 1894 she had earned a pilot's license and her own captain's license. Bowling, who succeeded his father as captain of the *Natchez*, was Blanche's mentor. "He taught me everything I know," Blanche claimed. "I would stand beside him at the wheel and repeat to him each snag, each bank, each plantation, each landing place.... I was constantly having my turn at the wheel, learning to take soundings, learning the signals, in fact all the intricate details that form part of a river captain's training.

"At times my husband was being called away, and breaking in a new man was always a troublesome process. So it was decided that I should apply for a captain's license." Blanche was at the helm of a new *Natchez,* her first command, when it steamed out of New Orleans in November 1894. She became something of a celebrity — and a character — along the Mississippi and became known as "the little captain." "I have done everything [on a steamboat] but marry people," she told a newspaper reporter. "You see, the captain has full authority on board over everyone, from the first mate to the cat. It is perhaps the only job in existence where there is no one to answer to. And if you own your own boat, you are answerable only to God."[6]

Blanche was the daughter of a Tensas Parish, Louisiana, cotton planter and had grown up beside the river. Her father was one of Tom Leathers' regular shippers and a friend as well. A newspaper reporter described Blanche: "A slight figure, five feet five inches in height, with the contour of charming womanhood, small, white, well-kept and perfectly molded hands ornamented with two handsome diamond rings; a face, grand, true, and ennobling to look upon, a fair skin glowing with the pink hue of health, perfect teeth, and a full red-lipped mouth that tells the story of a woman born to love, to feel, and to act kindly toward all humanity.... Captain Blanche is the angel of the Mississippi."[7]

Blanche's brother, Allan Douglass, was somewhat more straightforward in speaking of her: "Aw, she never did all that stuff, the actual running of the boat. Sure, she had licenses all right. But she let Bowling and the men do that. She tended to the buying and the service. She watched the bills and the cash. Maybe you call that running the boat after all."[8] Blanche, along with Bowling, retired from running steamboats in 1901. She died on January 25, 1940.

Mary M. Miller, Mary Becker Greene and Callie Leach French, all wives of Mississippi steamboat owner-captains, were other women who were licensed captains. Mary Miller received her license in 1884, making her perhaps the first woman captain and authorizing her to command steam vessels on the Mississippi, Red, Ouachita and other western rivers. She was married to George "Old Natural" Miller, who owned the steamboat *Saline*, which she also captained.

Callie French's husband was Augustus Byron French, who over a span of years operated five Mississippi River showboats, all of them named *New Sensation*. At least two of them operated at the same time, Augustus captaining one and Callie the other. Callie already held a pilot's license when she received her captain's license in 1895. Extremely versatile — and useful — she was much more than a steamboat captain and pilot. She wrote jokes for the showboat's players and on occasion even joined the troupe as one of the actors. When necessary, she cooked, sewed and did whatever nursing was required. Unlike Blanche Leathers, Callie — known as Aunt Callie to thousands who were the *New Sensation*'s patrons — did "all that stuff" that constituted the actual running of a steamboat. What's more, she did so without ever having an accident.

Mary Greene, known as Ma Greene on the river, received her captain's license in 1897 and with her husband, Gordon C. Greene, operated the *Bedford*, which they owned. She later captained the sidewheeler *Greenland* from Pittsburgh to St. Louis for the St. Louis World's Fair in 1904.

The captain's right-hand man in running a steamboat, the officer in charge of the nitty-gritty of boat operations, was the first mate, who on many steamers was the *only* mate. As the boat's second in command, he was responsible for running the boat when the captain was off duty and he did all that the captain did while in command. He was, though, first of all, boss of the deck crew, a job that often required him to be a stern disciplinarian and manager of men. Deckhands could be an unruly, willful, lazy, irresponsible and even mutinous lot, and it took a tough taskmaster to handle them. The veteran pilot Merrick described a mate who served with him and told how he handled his job. He was Billy Wilson, from Pennsylvania, smooth-shaven, red-faced, about five-foot-eight and about a hundred and sixty pounds, a well-read and ordinarily quiet man. The crew that Wilson managed comprised about forty deckhands, men hired off the riverfront at St. Louis, Galena, Dubuque and St. Paul, riffraff who would get drunk whenever they could get whiskey, which was not infrequent. It took Wilson's constant vigilance to prevent the deckhands from drinking too much and fighting among themselves, defying the boat's officers in the process. Wilson was, like other mates, a driver of men.

He made a habit of carrying with him a paddle fashioned from a barrel stave, with one end forming a handle and the other end flattened like a canoe paddle, into which had been bored holes a quarter inch in diameter. "When the case was one of mere sluggishness on the part of one of the hands," Merrick reported, "a light tap with the flat part of this instrument was enough to inspire activity. When the case was one of moroseness or incipient mutiny, the same flat side, applied by his [Wilson's] powerful muscles, with a quick, sharp stroke, would leave a blood blister for every hole in the paddle; and when a drunken riot was to be dealt with, the sharp edge of the paddle on a man's head left nothing more to be done with that man until he 'came to.' With a revolver in his left hand and his paddle in his right, he would jump into the middle of a gang of drunken, mutinous men, and striking right and left would intimidate or disable the crowd in less time than it takes to tell it."[9]

Over time, the deckhands who worked under Wilson learned to fear him. On a cold, rainy night on the upper Mississippi Wilson's boat put into a woodyard to take on fuel, and Wilson ordered the deck crew to turn out and carry the wood aboard, shouting, "All hands! Wood up!" Most turned out, but Wilson saw that ten or so hands were missing and went looking for them, paddle in hand. He first searched their bunks and failing to find them there, moved on to the boilers, under which, he knew, the men sometimes slept in cold weather, to keep warm and to dry out their clothes. He crawled under the boilers and started swinging his paddle blindly in the dark, striking whomever the paddle found, and issuing a loud stream of profanity as he did so. The shirkers quickly scooted out from under the boilers and fell into line with the other hands, carrying the wood aboard. At the next wooding stop, there was no one found soldiering, hiding out to escape work.

Wilson's technique was to speak softly and carry his big paddle — until provoked into strong language. Other mates drove the men with the force of their voices, backed up often by the force of their physique. Merrick recalled one, a second mate, a big man with a big voice, who "roared and swore at the crew all the time." Merrick claimed that in those days men who hired on as deckhands were at the very bottom of the social scale, men who, whether black or white, were used to being driven. "They would not work under any other form of authority," he said. It was the mate's duty to exercise the necessary authority, whatever it took to have the deck crew do their jobs, and mates were determined to do their duty. Eventually, however, after many reports of abuse of crewmen by steamboat officers, the United States Congress as well as some states, such as Louisiana, enacted legislation to regulate river navigation and define the powers of steamboat officers. Brutal treatment of crewmen was thus outlawed, and public outcry demanded the laws be enforced.

Handling some of the steamboat's more tranquil responsibilities was its chief clerk, who was in charge of the boat's business office, as a purser is today. He issued passenger tickets, received payments for passage and for the shipment of freight, kept the account books, made out the boat's payroll, answered passengers' questions, showed them around the boat, acted as cashier and generally made himself helpful to first-class passengers. The chief clerk usually had an assistant, officially designated the second clerk but commonly known as the mud clerk. He got the dirty work. One part of his job was to go ashore and oversee and record the receipt and delivery of freight, a duty that required working outdoors in all kinds of weather, including cold and rainy. Rain frequently turned earthen levees into muddy mounds, up and along which the clerk trudged to get his job done, after which he returned to the boat well muddied. Hence the term mud clerk.

In the middle of a gloomy night a lantern might be spied waving along the shore, signaling the steamer to stop for passengers or freight, and after the boat had pulled up to the riverbank, the mud clerk, by the light of iron torch baskets, filled with bright-burning pine wood or oily rags set alight and suspended from long poles aboard the vessel, would jump ashore to find out what business there was for the boat, while the mate shouted to the roustabouts to wake up and run out the landing stage to receive whatever was about to come aboard. When the boat needed refueling, it was the mud clerk who got the task of measuring the cords of wood taken aboard at wooding stops, to make sure the boat was getting all that it was paying for.

Along with its captain and pilots, a steamer's engineers — there were usually two aboard, the chief engineer and the second engineer — carried a certain prestige in the steamboat community. Engineering was considered a profession, and those who mastered it earned status — and a good deal of job security — among steamboat operators and fellow officers. Engineers ordinarily worked their way up to the position of chief engineer through years of learning the mechanics of propelling a steamboat. In the early days of steamboating engineers were not much more than engine tenders and were usually firemen who had graduated from stoking furnaces to running the boat's machinery, which at that time was not terribly complicated. As engines and mechanical systems became more sophisticated, though, more preparation for an engineer's job was demanded, and in order for an engineer to become licensed, considerable knowledge and skill were required.

The usual starting point for a steamboat engineer was as a striker, or cub, whose work was both taxing and constant. One of the most onerous chores of the beginning striker on the Mississippi was cleaning the accumulated mud from the boat's boilers, the thick sediment that had been unavoidably pumped into the boilers along with the river water. When the boat put into a port for a day or two, while other members of the crew were enjoying themselves ashore, the striker entered the boiler, from which the water had been drained, through a manhole, an opening in the boiler not much larger than the striker himself, and with a hammer and chain began pounding on the two flues and the sides of the boiler to dislodge the mud, which was then washed out of the boiler with a stream of water pumped through a hose. On a good-sized steamer there could be as many as eight boilers to be cleaned periodically.

Engineers were expected to be not only mechanics, able to repair the boat's machinery, but also skilled ironworkers, able to fabricate missing or broken parts of the machinery and make miscellaneous metal devices (including bolts, nuts, hog chains and chimney guy wires) from wrought-iron bars that were kept on board for such purposes. Steamboats were equipped with a blacksmith's forge and anvil, and the needed parts were beat out by the engineers with a twelve-pound striking hammer and a two-pound shaper.

Continuous maintenance was a large part of an engineer's job. The boat's boilers, its propulsion engines, its hoisting engine that was used for warping the boat over sandbars, and its doctor — the small engine used to pump water from the river into the boilers and into the boat's hoses — had to be repeatedly overhauled to keep them in perfect operating condition. Engineers went through a long checklist of things to do in preparation for each trip, so that when the signal came from the pilothouse to get under way, the big paddle wheels would actually start turning and the boat would begin moving.

The engineer received signals from the pilot by way of a set of gongs and jinglers, little tinkling bells that were attached to cords extending from the pilothouse down to the engine room. The pilot would pull on the appropriate bell cord to give a specific order to the engineer. Side-wheelers had a set of bells for each of its two engines. The boats also had a pipelike speaking tube that ran from the pilothouse down to the engine room on the main deck, ending in a flared opening, like the bell of a trumpet, mounted above the engineer's head. According to Mississippi River steamboat historians Herbert and Edward Quick, messages on the tube sometimes got a little heated: "When the signalers needed more expression than was afforded by the bells, the bells being unable to curse, no matter how furiously jingled, the speaking tube was used to good purpose." Engineers, however, were at a disadvantage when those tube exchanges took place, for the big end of the tube, in the engine room, was designed for listening, the small end, in the pilothouse, for speaking, making the conversation mostly one way. The engineer had a bell pull of his own, which he used to signal the firemen when he wanted them to open the furnace doors to help decrease steam pressure.

When the boat was under way, the engineer kept to his station in the engine room, despite fire, collision, grounding or some other emergency, for when the pilot rang the signals, the steamer's powerful engine, in the hands of the engineer, had to respond as ordered. "The engineer knew his job was as important as any aboard, even if he did do his work in grease and sweat and was usually laboring in obscurity while the captain and pilot were going through the pomp and show of landing and departing," the Quicks reported. "Let the captain play host and the pilot play lord.... But when she [the boat] drifted down to a shoal, and at just the right instant the pilot rang full steam ahead, it was the engineer on watch below who must know that his engines were ready, who must apply the power that carried her across."[10]

Not quite so dramatic was the role of the boat's steward. His main job was to keep the first-class passengers fed and comfortable. To help him do so he commanded a staff of assistant stewards, cooks, bakers, waiters, pantrymen and maids, among others, whom he both hired and supervised. Good stewards — and the cuisine they were able to produce for the benefit of passengers and a boat's reputation — were so valued by steamboat owners that their pay, about two hundred dollars a month, ordinarily equaled that of the chief engineer, the mate and the clerk and sometimes even that of the boat's captain, who made about three hundred dollars a month.

10 • *Owners and Officers* 141

The chief steward and two other officers aboard the steamer *J.M. White*. The steward's main job was to keep the first-class passengers well fed and comfortable with help from his staff of assistant stewards, cooks, bakers, waiters, pantrymen and maids, whom he hired and directed (Library of Congress).

It was a demanding job, though. The steward procured all necessary foodstuffs and therefore was required to be acquainted with all sorts of food suppliers in the river's port cities and obscure communities along the boat's route as well. He made up the daily menus, supervised the food preparation, oversaw all aspects of the dining service and directed the maids' care of the cabins and of the passengers' personal needs. He also planned and directed leisure activities for the passengers and to the passengers was the chief enforcer of the boat's rules. His guiding principle was, according to one account, to "spare neither pains nor money to make the passengers comfortable."

By the middle of the nineteenth century many, if not most, Mississippi River steamboat stewards were black men. It was a job at which they apparently excelled and one that rewarded them not only with handsome salaries but with status, particularly in the black communities along the river. Black stewards, however, sometimes ran into trouble with white workers whom they supervised and with white passengers who refused to accept the steward's authority, as was the case in one incident when a passenger seated himself at the officers' table and when told by the steward he was in violation of the rules, the passenger struck the steward in the head with his cane and told him, "No black bastard can tell me where to sit."

In a class of his own among steamboat officers was the pilot. "Pilots," the Quicks commented, "were treated with a great deal of respect by every one aboard a steamboat: captains, waiters, deckhands, clerks, engineers, firemen and passengers. Ashore they were envied for the pay they drew [about five hundred dollars a month, compared to the captain's three hundred].... They never asked any one for any thing: On all occasions they gave orders. They were, in fact, the accepted aristocrats of the steamboat business."[11]

To many youngsters growing up in towns along the Mississippi, the steamboat pilot was an object of envy and hero worship. He was the person that a boy with a fondness for the river would most like to be. Some who aspired to the position eventually made their dream come true, but only by dint of devotion to the study of the river and a dogged determination to master it. When the Mississippi was low, pilots rarely ventured downstream at night. The danger was too great. The boat at slow speed was hard to steer with the force of the current behind it, pushing it toward whatever obstacle lay before it. Voyages upstream in the dark were somewhat safer, because of better steering against the current. It was important, furthermore, that nothing interfere with the pilot's vision at night, and pains were taken to prevent any diminution of his ability to see as best he might in the dark. Canvas curtains were draped over the forward part of the boiler deck, where the cabin passengers were, and the forecastle was covered to block the light emanating from the furnace doors. Skylights were also draped, and according to one account, on the hurricane deck, up near the pilothouse, not even lighted cigars were allowed.

Feeling their way along the river at night, pilots sometimes made use of not only what they could see but what they could hear. A pilot on the Missouri River took notice of a well known dangerous spot by the sound of a barking dog, which always came from the same location along the river, and he blindly steered the boat to avoid the danger by the sound of the barking. One night, though, the dog barked from a different place, and shortly thereafter, the boat ran onto a sandbar and was wrecked.

The fund of knowledge required of a pilot was enormous, as the Quicks pointed out. "The depth of the channel at every point, at all stages of the river, the amount of water on the bars, the habits of the river in tearing down banks and building up bars, the changing depth of the water between one trip and the next — all these things he had to know; also the marks, the trees, houses, points, bars, posts, hollows — anything by which to lay his course through every mile of the river. And as the river changed, he had to know the new marks, new channels, new bars, new islands."[12]

Pilots had to know not only what was *along* the river but what was *in* the river. They had to know how to read the river. A line appearing on the river's surface, for example, told the pilot, peering ahead through the windows of the pilothouse, that a reef of mud and sand lay just beneath the surface. The wind sometimes made a similar mark on the river, but not exactly, and the pilot had to know the difference. A combination of lines and circles on the water revealed a spot that was developing into a shoal. A dimple on the water indicated a rock below. A streak like a boat's wake showed that a snag lay hidden right below the surface.

At times when it was simply impossible to see or otherwise know what lay ahead, as frequently happened on the lower Mississippi when sugar cane growers burned the cane stalks after the juice had been pressed from them and smoke from the burning heaps of cane spread over the river like a dense fog, prudent pilots put their boats into shore and tied up against the levee until daylight came or the wind shifted and removed the smoke from the river.

"You can not steer a boat by landmarks ten feet ahead of her," former pilot Merrick remarked. "The pilot searches for landmarks a mile away, and must be able to distinguish between two kinds of blackness — the blackness of the night below [on the river], and the blackness of the sky above, and from the dividing line between the two must read his marks and determine his course."[13]

Thomas Burns, a native of Boston who grew up in Galena, Illinois, was a pilot on the upper Mississippi just before and right after the Civil War and knew the upper Mississippi so well, it was said, that he could get out of bed on the darkest night, head for the pilothouse to take over his duties and before he reached the pilothouse door could tell in what section of the river the boat was steaming.

For cub pilots, Burns had a guiding principle to help keep them and their boats out of serious trouble in time of unknown danger. Having learned that a steamboat standing still in the water was less likely to damage itself or inflict damage on any other boat in the stream, he would tell his student pilots, "When in doubt, ring the stopping bell and set her back." A memorable and merciful golden rule.

By the 1880s, when Samuel Clemens revisited the Mississippi after a twenty-one-year absence, electricity and navigational aids had reduced the demands on steamboat pilots, as Clemens discovered on his voyage up the river. "As we approached famous and formidable Plum Point," he wrote, "darkness fell, but that was nothing to shudder about — in these modern times. For now the national government has turned the Mississippi into a sort of two-thousand-mile torchlight procession. In the head of every crossing, and in the foot of every crossing, the government has set up a clear-burning lamp. You are never entirely in the dark now; there is always a beacon in sight, either before you, or behind you, or abreast."[14]

Clemens conceded that lighting the Mississippi's danger spots had made river travel safer and made the pilot's job easier, but the improvement had come at a cost. "This thing," he complained, "has knocked the romance out of piloting."

Steamboat officers alternated their duties on shifts, or watches. The captain and the mate were on duty six hours and off duty six hours, one relieving the other, around the clock. The chief clerk and the mud clerk also were on six hours and off six. The engineers and pilots generally stood a three-hour watch, believed to lessen the monotony of the job and the boredom of off-duty time. The captain and the chief clerk began their first watch of the day at six A.M., right after breakfast. At noon they were relieved by the mate and the second clerk, who were on duty till six P.M., at which time the captain and chief clerk returned to duty. After supper, the mate and second clerk were able to sleep until midnight, when they had to rouse themselves and go back on watch.

"While each class of officers was on duty the same number of hours each day," George Merrick pointed out, "the difference lay in the fact that the junior officers were compelled by this arrangement to turn out at midnight throughout the season. It was this turning out at midnight that made the mate's watch ... very undesirable.... A man can knock about until midnight very agreeably, after a short nap in the afternoon, provided he can have a sound sleep during the 'dead hours' from midnight until six o'clock in the morning. To turn out at midnight every night and work until six is an entirely different matter."[15]

• 11 •

The Perils

On the evening of May 4, 1825, the steamer *Teche* pushed away from the wharf at Natchez, bound for New Orleans, heavily laden with bales of cotton and carrying some seventy passengers, many of whom had boarded the boat at Natchez. Darkness, worsened by a thick haze, quickly descended on the river, and by the time the *Teche* had traveled but ten miles the boat's captain decided it was unsafe to proceed farther. He ordered the boat anchored in the stream until conditions improved. About two o'clock the next morning, the haze having dissipated, the anchor was hauled up, and with its steam already raised, the *Teche* began to move ahead again. Suddenly an explosion that sounded like an artillery bombardment shook the boat, jarring the sleeping passengers from their berths.

All lights aboard the boat immediately went out, extinguished either by the escape of steam or the concussion of the air. In the darkness that instantly followed, a frightened crowd of passengers assembled on the deck, unaware of exactly what sort of disaster had struck them. Then swiftly came a shout that the boat was on fire, and the crowd broke into pandemonium, the panicked passengers rushing helter-skelter about the deck in the darkness, some leaping overboard to escape the flames.

The exact number of lives lost in the *Teche* disaster was never determined, but several persons were known to have been immediately killed by the explosion, and others were so severely scalded or otherwise injured that they died not long afterwards. No fewer than twenty, and perhaps as many as thirty, persons drowned.

On August 12, 1828, the steamboat *Grampus*, carrying passengers and towing three sailing brigs and a sloop up the Mississippi to New Orleans, was rocked by an explosion that blasted the captain and a passenger from the wheelhouse and landed them fifty feet away on the forward deck, severely bruised but alive and whole, surrounded by debris. The pilot who had been standing at the wheel was thrown into the river and drowned. The boat's other pilot, who had been walking on the deck just outside the wheelhouse, suffered a broken leg and other injuries and subsequently died. The brig in tow on the larboard side of the *Grampus* had both topmasts cut off by flying fragments of the *Grampus*'s machinery, and the brig being towed on the steamer's starboard side had her bottom penetrated by a piece of the *Grampus*'s boiler.

Altogether, nine persons lost their lives, some killed instantly by the blast, some who died later of their injuries. Four others survived.

The *Grampus* explosion was determined to have been the fault of human rather than mechanical failure. The assistant engineer who was in charge of the engine room while the chief engineer was off duty had fallen asleep after partly shutting off the water supply to the boiler, which soon overheated. Awaking, the assistant engineer quickly noticed that gauges showed the water at a dangerously low level in the boiler and immediately turned on the pump to resupply water to the boiler. When the new water entered the white-hot boiler, it was instantly turned into steam, creating an excess of pressure that burst the boiler.

On February 24, 1830, one or more boilers of the steamer *Helen McGregor*, on its way from New Orleans to Louisville, exploded while the boat was tied up at the wharf in Memphis. A section of deck near the boilers was crowded with people, all of whom were either killed or injured. As many as sixty persons were believed to have died from the blast, including an unknown number whose bodies were hurled into the river and never recovered. The *Helen McGregor* explosion was the deadliest in the history of steamboats up till that time.

On June 9, 1836, near Columbia, Arkansas, the *Rob Roy*, en route from New Orleans to Louisville, stopped its engine long enough for the engineer to oil part of its machinery, and in the two minutes or so that the engine was stopped, the steam in the boiler accumulated so rapidly that it burst the boiler. Immediately following the blast, the boat was run ashore, allowing passengers and crew to escape lest they drown as the boat burned and sank. The only lives lost were those of the victims of the explosion itself.

Just after 5 P.M. on November 15, 1849, the *Louisiana*, an elegant new steamboat captained by John Cannon, was backing out of its berth at the foot of Gravier Street in New Orleans, two blocks from Canal Street in the heart of the business district, when a horrific blast shattered it and blew the superstructure off the two steamers on either side of it. Human bodies, persons who had been aboard the *Louisiana*, were blown two hundred feet into the air, one of them flew like a projectile through the pilothouse of the *Bostona*, one of the steamers docked beside the *Louisiana*. People standing as far as two hundred yards from the boat were struck by flying debris. With little time for its passengers to escape to shore, those who had survived the explosion, the broken remains of the *Louisiana* quickly sank into the river. An estimated eighty-six persons lost their lives, including several who were aboard the two boats beside the *Louisiana*.

Captain Cannon and his chief engineer, John L. Smith, were ashore on business at the time of the explosion and came in for much criticism for their absence. The coroner's jury that investigated the disaster determined that the boat's second engineer, Clinton Smith, was "grossly, culpably, and criminally neglectful and careless of his duty." The jury also reported that Cannon was "highly culpable" in allowing Clinton Smith to be in charge of the engine room and blamed Cannon and his chief engineer as the "cause and causes of said explosion." In a later hearing, however, both Cannon and the chief engineer were acquitted of a charge of manslaughter in connection with the accident.

Starting when the boiler on Henry Shreve's *Washington* blew apart in June 1816, the list of steamboat explosions was long and grim. An estimated seventy-six steamboat explosions on the Mississippi and its tributaries occurred between 1836 and 1848, a rate of about one every six months. In the years 1846 to 1848 steamboat explosions on the western waters were reported to have caused 259 fatalities.

As destructive and deadly as explosions were, fire was the most fearsome danger. In the early-morning darkness on May 8, 1837, the *Ben Sherrod*, running between New Orleans and Louisville and racing with the steamer *Prairie*, fell victim to one of the Mississippi's worst steamboat fires. About one o'clock in the morning the boat was about fourteen miles above Fort Adams, Mississippi, racing to pass the *Prairie*, just ahead of it. The firemen were shoving in pine knots and sprinkling rosin over the coal, doing their best to raise more steam. The boilers became so hot that they set fire to the sixty cords of wood on board, and the *Ben Sherrod* was soon completely enveloped in flames.

The boat's yawl was finally launched, but it was so overloaded with fleeing crewmen that it sank, and nearly everyone in it drowned. Passengers desperate to escape the fire leaped into the river, still in their night clothes. Ten women passengers jumped overboard, some of them quickly drowning once in the water, others finding floating debris to which they could cling. Only two of the ten survived.

The *Prairie*, which the *Ben Sherrod* was trying to overtake, continued on up the river without stopping or turning about to try to save those aboard the stricken vessel. When the *Prairie* reached Natchez, its captain reported

The *Ben Sherrod* (the name misspelled by the artist) ablaze on the Mississippi, near Fort Adams, Mississippi. The boat was racing the *Prairie* when its boilers overheated and set off a fire that consumed the vessel in the early-morning darkness of May 8, 1837. Many of its passengers and crew drowned after leaping into the river to escape the flames (Library of Congress).

that the *Ben Sherrod* had caught fire and at Vicksburg he made a similar report, for all the good it did.

Another steamer, the *Alton*, which reached the burning boat within half an hour after the fire started, failed to stop to give aid and actually contributed to the accident's death toll. The *Alton* came steaming up on the scene, amid the exhausted survivors in the water, holding onto floating debris, and the turbulence caused by its paddle wheel sucked several survivors under water and drowned them. A man holding onto a floating barrel and helping a woman hold on to it, too, was washed underwater by the *Alton*, as was the woman. The woman drowned; the man bobbed up to the surface and floated fifteen miles downriver before being rescued by the steamer *Statesman*. A man named McDowell and his wife were both in the river when the *Alton* arrived. He managed to stay afloat and was swept by the current two miles downriver, where he then swam ashore. His wife, though, holding onto a wooden plank, was pulled under by the *Alton* and drowned. McDowell's son also died in the disaster.

Survivors told other stories of horror. A young wife and mother, Mary Ann Walker, awakened by shouts of "Fire!" dashed out of the women's cabin holding her infant child, trying to reach her husband. Unable to get to him in time, she watched as he fell into the flames and she then jumped into the river to save herself and her child. She grasped a plank and was within forty yards of being rescued by the *Columbus*, another steamer that had come upon the tragic scene, when she suddenly sank out of sight and was seen no more. A young man who had fled to the hurricane deck and escaped the flames turned back to the blazing cabin when he heard his sister's cries. Trying to save her, he clasped her in his arms as the flames overtook them. Both burned to death.

One of the *Ben Sherrod*'s clerks, one of its pilots and its mate all burned to death, as did all the boat's chambermaids. Of the thirty-five negro crewmen on the vessel, only two survived. Among the dead were two of the captain's children and his father. The captain's wife, one of the ten women who leaped into the water, survived but was severely burned. The captain also survived. Not more than six or seven passengers survived. The charred wreck of the *Ben Sherrod* sank beneath the Mississippi's dark waters just above Fort Adams.

The steamboat *Brandywine* left New Orleans in the evening of April 3, 1832, bound for Louisville, carrying some two hundred and thirty passengers as well as freight, including a number of carriage wheels packed in straw and stacked on the boiler deck, near the officers' cabins. Sometime during its voyage the *Brandywine* became engaged in a race with the steamer *Hudson* and fell behind when it was forced to stop for repairs. Back in the race following the repairs, the *Brandywine* attempted to gain speed and make up its lost time by feeding more rosin into furnace, thereby intensifying the heat and increasing steam pressure. About thirty miles above Memphis, about seven o'clock in the evening of April 9, the *Brandywine*'s pilot, in the pilothouse, noticed that the carriage wheels' straw packing was on fire and quickly gave an alarm.

The captain and crewmen immediately responded, trying desperately to extinguish the flames and pulling out the burning wheels and throwing them overboard. Their efforts, however, only exacerbated the blaze by allowing the wind to whip through the separated mass of straw-packed wheels, spreading the flames to other parts of the boat. In less than five minutes after the pilot had given the alarm, the entire vessel was ablaze. The boat's yawl quickly filled with frightened passengers and was lowered into the water. No sooner had it touched the stream than it overturned and sank, leaving the passengers to the mercy of the Mississippi.

The pilot steered the flaming steamer, still under way, toward shore, hoping to beach it. About a quarter of a mile from the riverbank the boat ran aground and stuck fast on a sandbar in nine feet of water. Passengers and crewmen still aboard either perished in the flames or hurled themselves into the river and tried to swim to shore. Of the two hundred and thirty passengers the *Brandywine* carried, an estimated seventy-five survived the disaster. The rest either burned to death or drowned.

The *Belle of Missouri* was far more fortunate. En route from New Orleans to St. Louis, it stopped just above Liberty, Illinois, to take on wood and caught fire while docked. Its two hundred or so passengers fled safely to shore—and none too soon, for a shipment of gunpowder aboard the boat exploded not long after their escape.

The *Clarksville* caught fire near Ozark Island in the Mississippi on May 27, 1848, and its pilot promptly turned for shore. Just as the bow of the boat struck the riverbank, flames broke into the main cabin, one of the boilers exploded and, simultaneously, three barrels of gunpowder ignited, creating a huge cloud of black smoke. The captain, named Holmes, jumped overboard with his wife, left her on shore, then returned to the stricken vessel to direct the evacuation of the other passengers, shouting to them, "Pick up chairs, everybody! Jump overboard. But take your chairs. They'll give you something to hold you up!" When the last person was off the boat, Captain Holmes, suffocating from the pall of smoke, leaped from the upper deck. His body struck the railing on the lower deck, and he was thrown into the flames and burned to death. All of the cabin passengers survived, although many were injured, including the governor of Tennessee. Thirty deck passengers, however, at the stern of the boat and evidently too frightened to dash through the smoke to the bow and jump overboard, lost their lives.

On October 8, 1849, five steamboats — the *Falcon*, the *Illinois*, the *Aaron Hart*, the *Marshal Ney* and the *North America*— that were docked at the Poydras Street wharf in New Orleans caught fire and burned as flames spread from boat to boat. Two other steamers in nearby berths managed to back into midstream and escape the inferno with only minor damage.

The *Martha Washington*, on its way from Cincinnati to New Orleans, caught fire in the Mississippi at one-thirty in the morning on January 14, 1852. Within three minutes the boat was engulfed by flames, blazing from stem to stern. Only a few of the passengers were lost, however, and only one of the crew, the boat's carpenter.

In the steamboat's earliest days explosions and fires could be attributed to the crudeness of the propulsion systems, the boilers in particular. It took time for manufacturers and engineers, advancing the steamboat's machinery largely through trial and error, to learn how to build safe boilers and have them used safely. But as steamers became commonplace and water transportation became the major mover of freight and passengers in the Mississippi valley, competition overtook the concern for safety. Speed became the first objective of steamboat owners and operators. Attempts to beat old records and outrace the competition led to abuses of the machinery and, in many cases, the complete abandonment of caution. Captains would order their boats' fireboxes crammed with fuel, and fires were made to burn ever hotter by adding to the flames pitch, rosin, oil or pork fat — then tying or weighting down the automatic safety valves on their boilers to raise steam pressure to an explosively high level.

At the same time that newspapers were reporting the latest disasters on the river they were also editorializing against the dangers presented by owners and operators who put speed above safety, widely believed to be the ultimate cause of most steamboat explosions and fires. Letters to the editor further decried the dangerous practices. "Want to know why boilers bust on leaving shore?" one former steamboat captain wrote in a letter published in the New Orleans *Picayune* in November 1840. "Steamboat men and even passengers have a pride in making a display of speed. To do this they hold on to, instead of letting off, steam. The flue gets hot and the water low, and the first revolution brings the two elements in contact and causes a collapse."[1]

Another reason for the alarming number of fiery disasters was thought to be simple disregard of the hazards of traveling with combustible materials aboard a wooden boat. An appalled Frenchman visiting in the United States, Michael Chevalier, wrote that "Americans show a singular indifference in regard to fires. They smoke without the least concern in the midst of half-open cotton bales, with which a boat is loaded; they ship gunpowder with no more precaution than if it were so much maize or salt pork, and leave objects packed in straw right in the torrent of sparks that issue from the chimneys."[2] Like other critics, Chevalier also noticed that speed trumped all safety considerations aboard the steamers.

Another deadly peril on the river was boat collisions, one of the worst of which involved the steamer *Monmouth*. It left New Orleans on October 23, 1837, headed for the Arkansas River, carrying 611 Creek Indians to a reservation where they were to be resettled. On the night of October 30, a particularly dark night, the *Monmouth* was steaming through a part of the Mississippi known as Prophet Island Bend and there encountered the ship *Tremont*,[3] which was being towed down the river by the steamboat *Warren*, obscured by the darkness and evidently unseen by the *Monmouth*'s officers until the last minute. In a desperate effort to avoid the oncoming *Tremont*, the *Monmouth* apparently swerved, but too late. The prow of the *Tremont* caught the *Monmouth* broadsides, smashing into the steamer with such an impact that the *Monmouth*'s main cabin was separated from its hull. The hull sank almost immediately, but the cabin was sent drifting downstream on the current until it broke in two, spilling all of the *Monmouth*'s passengers into the night-shrouded river.

The crewmen of the *Warren* and those of another steamer that arrived on the scene, the *Yazoo*, managed to save about three hundred of the *Monmouth*'s passengers from the river. The rest drowned. Also lost were two of the *Monmouth*'s crew, the fireman and the bartender.

Blame for the collision was placed on the officers of the *Monmouth*, who failed to observe the Mississippi's rules of the road. "This boat [the *Monmouth*]," according to a nineteenth-century account, "was running in a part of the river where, by the usages of the river and the rules adopted for the better regulation of steam navigation on the Mississippi, she had no right to go, and where, of course, the descending vessels did not expect to meet with any boat coming in an opposite direction."[4]

On November 19, 1847, the steamer *Talisman* was approaching Cape Girardeau, Missouri, when it was rammed by the *Tempest*, which struck it just forward of its boilers. The *Tempest* backed away, exposing an enormous hole in the *Talisman*'s side, through which water was rapidly pouring. In the *Talisman*'s pilothouse the pilot was furiously ringing the engine room bells to order more speed as he headed the vessel for the riverbank, trying to reach it before the boat, quickly sinking, slipped beneath the surface. The engine room meanwhile was filling with water. The chief engineer, named Butler, ordered his strikers to get out of the engine room and seek safety, but refusing orders from the pilothouse to also leave, he remained at his post to keep the crippled steamer under way as long as possible. In less than ten minutes the engine room filled with water, and the *Talisman* went down, taking chief engineer Butler with it to the bottom of the river. The *Tempest* stood by to help rescue the *Talisman*'s passengers and crew, but despite its efforts, more than fifty of those aboard the *Talisman* lost their lives.

The *Archer*, operating out of St. Louis, was struck by the steamer *Di Vernon* five miles above the mouth of the Illinois River on November 27, 1851, and was cut in two. It sank in three minutes, taking forty-one lives.

A twentieth-century writer gave some understanding of the problem of collisions on the river, particularly collisions involving towboats, which tow barges or other vessels by pushing them, making of them an unwieldy burden:

> When you come to consider it, there should be more collisions than there are on the river. Especially on some of the smaller tributaries, the bends are so sharp that there is no way to see around them; and the hills go up on both sides, hiding any trace of smoke that may warn a pilot of another tow. You might hear the other fellow's whistle, and then again you might not. It depends on the wind

and the noises on your own boat.... The emergencies on the river are awful in their slowness. You see them a long time ahead and you do what you can; and if that's not good enough, you'll wish you hadn't ever come on the river, and you wait for the crash — ten or fifteen minutes with no way out, no way to stop it from happening.[5]

Other vessels, though, were not the only potentially disastrous hazards that steamers encountered on the river. On January 3, 1844, the *Shepherdess* was ascending the Mississippi on its way to St. Louis from Cincinnati, steaming into the stormy winter night through frigid water, most of its seventy or so passengers asleep, though several in the men's cabin chose to huddle around the stove to keep warm. Around eleven o'clock, without any warning, the boat plowed into a snag — an obstruction in the river, usually a fallen tree — near Cahokia, Illinois, ramming it with such force that several planks were torn from the forward part of the boat's hull. Water instantly began rushing into the gaping hole and in less than two minutes the water had risen to the lower deck. The captain, A. Howell of Covington, Kentucky, who had recently bought the *Shepherdess* and was making his first trip on it, ran to the women's cabin to reassure the ladies, telling them there was no danger, then returned to the forecastle, which was awash as the bow of the vessel was slipping beneath the surface. He later was apparently swept overboard by the rising water and drowned.

Within three minutes the water had reached the upper deck, and passengers there could see from the stern railing people in the river, struggling to stay afloat in the icy stream. Some passengers on the lower deck had been able to save themselves by climbing into the yawl, which they cut loose as the water rose and, finding no oars in it, paddled it to shore with a broom. The only safe place left aboard the vessel was the hurricane deck, but it became difficult to reach as the boat's bow sank deeper into the river. The only way to get to it was from the stern. Most, if not all, of the passengers that had been in the main cabin, on the boiler deck, managed to reach the roof of the hurricane deck. Some of them were aided by passenger Robert Bullock, a young man from Maysville, Kentucky, who with little regard for his own safety went from stateroom to stateroom and whenever he heard a young child crying he took the child and handed him or her up to someone on the hurricane deck. Among those he helped save was the so-called Ohio Fat Girl, a 440-pound woman who was a member of a carnival troupe.

The powerless vessel, carried downstream by the current, crashed into a second snag, this rising one above the surface. The boat nearly capsized when it struck the snag, tipping over on its larboard side, then lurching to starboard, and spilling a number of passengers into the icy river, only some of whom were able to swim to shore. The boat righted itself and continued to drift downstream, then struck the riverbank so hard that the cabin was separated from the hull. The hull then sank on a sandbar; the cabin continued a short distance and it, too, hit a sandbar and became stuck.

The steamer *Henry Bry* had stopped at Carondelet, just below St. Louis, and as the cabin of the *Shepherdess* floated past its position, the *Henry Bry*'s captain ordered his yawl launched into the river to rescue as many as he and his crewmen could by making repeated trips to haul survivors to safety. The ferryboat *Icelander* came down from St. Louis to join the rescue effort around three A.M. and removed all the remaining survivors from the *Shepherdess*'s marooned cabin. An estimated forty persons, many of them young children, failed to survive the disaster.

Snags and other objects in the river had been a menace ever since the first steamboat on the Mississippi, the *New Orleans*, had its hull pierced by a stump and sank in 1814. Another early victim was the *Tennessee*. Steaming upriver through a snowstorm, its pilothouse windows coated with snow and its pilot unable to see through them, it ran into a snag near Natchez on the night of February 8, 1823, and had a hole the size of a door torn in its hull. Its yawl was lowered into the water and it took one load of passengers to shore, but with only one oar to propel it, it made just one trip. Many of the passengers and crewmen who were left aboard jumped into the river when they felt the boat going down. Some found floating wreckage to cling to until they could be rescued by skiffs that were rowed out from shore to save them. The rest were not so fortunate. Sixty persons were lost to the river.

The steamer *John L. Avery* left New Orleans on March 7, 1854, and about forty miles below Natchez on March 9, while it was apparently racing another steamer, it struck what was believed to be a tree that had been washed into the river by a recent rain. Water immediately rushed into the boat's hold through the pierced hull. The boat's carpenter and J.V. Guthrie, one of the engineers, were standing just forward of the boilers when the crash occurred, and the carpenter dashed to the hold to assay the damage, but the water was pouring in too fast to do anything about the leak, and the carpenter had to quickly retreat back to the deck. Guthrie then hurried for the engine room, but the water was up to his knees before he reached it. The cabin passengers quickly sought refuge on the hurricane deck. Minutes later the hull separated from the cabin and went down in sixty feet of water.

Six persons who had remained in the main cabin were rescued by the captain, J.L. Robertson, and the boat's two clerks, who lifted them from the rising water through a skylight onto the hurricane deck. One of those pulled through the skylight was a woman with one of her children her arms. She

11 • The Perils   155

The ice gorge that trapped five steamers in the Mississippi between Cairo, Illinois, and Columbus, Kentucky, in February 1872. Along with explosions, fires, collisions and snags in the river, ice, which could rip open a hull and sink a vessel, was one of the perennial perils for steamers on the upper Mississippi (Library of Congress).

had to be restrained from plunging back into the cabin, nearly filled with water, to rescue her baby, who had been asleep on the woman's bed. Many of the deck passengers were trapped by freight on the main deck and drowned as the boat sank. The second mate and another person launched the steamer's yawl, but it was almost immediately turned over by panicked passengers fleeing the doomed vessel. Twelve of the boat's twenty firemen drowned. Witnesses told of seeing struggling passengers and crewmen in the river, going down one by one in the murky water. At least eighty persons lost their lives in the *John L. Avery* disaster.

On the upper Mississippi ice was a hazard that claimed several steamers, including the *Iron City*, crushed by ice in the river at St. Louis on December 31, 1849; the *Northern Light*, which struck ice along the shore about 1862; the *Fanny Harris*, sunk by ice in 1863; and the *Metropolitan*, sunk by ice at St. Louis on December 16, 1865. Rocks were another peril. The *J.M. Mason* struck a rock that sank it in 1852, and the *H.T. Yeatman* ran into rocks and sank at Hastings, Minnesota, in April 1857. Sandbars were often no worse than a hindrance when steamboats ran onto them, but sometimes they would sink a boat, as one did to the *Kentucky No. 2*, which sank on a sandbar near Prescott, Wisconsin. In some cases the boat that had hit a sandbar was not sunk but simply left hopelessly stranded, unable to be lifted or pulled off, doomed to become a derelict after its passengers and crew had been removed from it.

One of the oddest river hazards was encountered by the steamer *Baltimore*. In 1859 it had the misfortune of running onto the underwater wreck of the steamboat *Badger State*, which had sunk three years earlier. With its hull opened up by the submerged vessel's superstructure, the *Baltimore* quickly went down, so fast that it settled atop the hulk of the *Badger State*.

Bridges were — and still are — a potential hazard to steamboat navigation. The captain of the modern excursion boat *Mississippi Queen*, replying to a passenger's question about the worst danger he faced on the river, said it was bridges that presented the biggest concern. If the boat should strike a bridge at a bad angle, there was a great danger that the boat, top heavy as it is, would capsize. The danger posed by bridges has existed since 1856, when the first bridge across the Mississippi River was completed after three years of construction. It was a railroad drawbridge that spanned the river between Rock Island, Illinois, and Davenport, Iowa. It consisted of three parts — a bridge over what is a narrow channel that passes between the Illinois shore and an island in the river (Rock Island), a section of track that ran across the island, and a long bridge over the wider part of the river, between the island and the Iowa shore. The placement of the bridge was disadvantageous for boats, for at that spot, there were cross-currents and boils produced by the chain of rocks in the river, all of which made boat navigation there a pilot's challenge.

When it was proposed, the bridge met powerful opposition from steamboat interests, who argued that it would be a hazard to navigation, and from the United States secretary of war, Jefferson Davis, who issued a ruling prohibiting its construction on the grounds that it would cross a government reservation. Nevertheless, it was built, the railroad interests proving more powerful than those who opposed the bridge.

On the morning of May 6, 1856, fourteen days after the first train had chugged across the newly completed bridge, the steamer *Effie Afton* passed under the bridge's draw and when it was some two hundred feet above the bridge, its starboard paddle wheel inexplicably stopped, and the boat turned, swung back and crashed into one of the bridge's piers. The impact apparently knocked over the stove in the *Effie Afton*'s galley, starting a fire that rapidly spread over the boat, destroying it and a section of the bridge as well. A number of lives were lost.

The *Effie Afton*'s owners swiftly sued the bridge company for damages. Suspecting that sympathy for the steamboat owners and for the *Effie Afton*'s victims would be strong, the directors of the bridge company carefully chose a lawyer who could persuasively argue their case. The man they picked was a forty-seven-year-old lawyer from Sangamon County, Illinois, who had been recommended to them as "one of the best men to state a case forcibly and convincingly, with a personality to appeal to any judge or jury hereabouts." His name was Abraham Lincoln.

The case went on the docket as Hurd vs. Railroad Bridge Company and was tried before Justice John McLean in Circuit Court in September 1857. Lawyers for the steamboat company presented two main arguments. One, that the Mississippi River was the great waterway for the commerce of the entire Mississippi valley and it could not be legally obstructed by a bridge. And two, the bridge involved in the accident was so situated in the river's channel at that point that it constituted a peril to all water craft navigating the river and it formed an unnecessary obstruction to navigation.

Lincoln argued that "one man had as good a right to cross a river as another had to sail up or down it." He asserted that those rights were equal and mutual rights that must be exercised so as not to interfere with each other, like the right to cross a street or highway and the right to pass along it. From the assertion of the right to cross the river, Lincoln moved on to the means of crossing it. Must it always be by canoe or ferry? he asked rhetorically. Must the products of the vast fertile country west of the Mississippi be forever required to stop at the west bank of the river, there to be unloaded from railway cars and transferred to boats, then reloaded onto cars on the east side of the river? The steamboat interests, he argued, ought not to be able to so hinder the nation's commerce and to stifle the development of the extensive area of the country that lay west of the Mississippi.

Lincoln conceded that the currents at the site of the bridge were problematic, but on the fourteenth day of the trial he presented a model of the *Effie Afton* and used it to show the jury, with the support of witnesses' testimony, how the accident occurred and to contend that it was the fault of those in command of the *Effie Afton*, that the boat had altered its course from the safe channel in order to pass another steamer, the *Carson*, which was also ascending the river. "The plaintiffs have to establish that the bridge is a material obstruction," he declared in his closing argument, "and that they have managed their boat with reasonably care and skill."

The jury apparently did not think the *Effie Afton*'s owners had made the case Lincoln said they must. The jury failed to agree on a verdict and was discharged.

On May 7, 1858, a St. Louis steamboat owner, James Ward, filed a petition in the U.S. District Court of the Southern Division of Iowa asking that the bridge be declared a nuisance and that the court order it removed. The federal district judge, John M. Love, so ordered, but on appeal to the U.S. Supreme Court in December 1862 the order was overturned and the bridge was allowed to remain. Complaints about the bridge continued to come from steamboat owners and captains, to the point that in 1866 the United States Congress passed an act requiring the Rock Island bridge to be dismantled and replaced by a new one farther up the river, the cost of which was to be shared by the railroad and the United States government. The new bridge, erected just above Rock Island, was completed in 1872, and the old one was torn down.

Bridge laws enacted by Congress in the 1870s allowed railroads to build bridges at locations specified in the enactments, but the bridges were required to be of a design and style of construction that would avoid serious obstructions to river navigation.

As for the other steamboat perils, a spokesman for the steamboat industry, William C. Reffield, evidently seeking to avoid further regulation, assured the United States secretary of the treasury, "The magnitude and extent of the danger to which passengers in steamboats are exposed, though sufficiently appalling, is comparatively much less than in other modes of transit with which the public have been long familiar.... It will be understood that I allude to the dangers of ordinary navigation [on the seas], and land conveyances by animal power of wheel carriages. In the former case, the whole or greater part of both passengers and crew are frequently lost, and sometimes by the culpable ignorance or folly of the officers in charge, while no one thinks of urging a legislative remedy for this too common catastrophe. In the latter class of cases, should inquiry be made for the number of casualties occurring ... and the results fairly applied to our whole population and travel, the comparatively small number injured or destroyed in steamboats would be a matter of great surprise."[6]

Even so, the steamboat trade publication *Lloyd's Steamboat Directory* in 1856 listed eighty-seven major disasters that had occurred on western rivers up to that time, many of which had each claimed a hundred or more lives. Also listed were 220 "minor disasters," as the publication called them. Between 1811, when the first steamboat descended the Mississippi, and 1850, at the midpoint of the golden age of the Mississippi River steamboat, steamboat accidents on the Mississippi killed or injured more than four thousand people.[7]

11 • The Perils

By far the worst catastrophe on the Mississippi, the deadliest maritime disaster in the nation's history to this day, taking more lives than did the *Titanic*, was the tragic loss of the steamer *Sultana*. It was a large — two hundred and sixty feet long, forty-two feet in the beam — handsome side-wheeler that had been launched at Cincinnati on January 3, 1863. It had been built for Captain Preston Lodwick, owner of three other steamboats, who paid $60,000 for it and who within the first year of the boat's operation had made more than twice that amount carrying freight and Union troops under contracts with the U.S. government.

It was equipped with four tubular boilers arranged horizontally and parallel to one another. The boilers were of a kind rarely used on the muddy Mississippi, because the mud carried by the river water, which was drawn into the boilers to produce steam, accumulated not in mud drums but in the two dozen five-inch-wide flues (or tubes) inside each boiler, necessitating frequent cleaning of the flues to prevent them from clogging and bursting.

The *Sultana* was built to accommodate seventy-six cabin passengers — the first-class passengers who occupied the staterooms — and three hundred deck passengers. Its legal capacity, then, was three hundred and seventy-six, plus the eighty to eighty-five members of the crew.

The boat's captain was thirty-four-year-old Cass Mason, who with two partners had bought the *Sultana* from Lodwick for $80,000 on March 7, 1864. Mason had first become a captain after marrying Mary Rowena Dozier, daughter of Captain James Dozier, who owned several steamers, including one named *Rowena*, which Dozier gave Mason to command. Mason apparently decided to use his new position to make a little money on the side by dealing in contraband, at which he was caught red-handed. On a trip downriver the *Rowena* was stopped by a Union gunboat on February 13, 1863, near Island No. 10 in the Mississippi and had its cargo searched. The search revealed a quantity of quinine destined for Tiptonville, Tennessee, then held by Confederate forces, and three thousand pairs of Confederate uniform pants. The gunboat's commander confiscated the cargo and the *Rowena*, which the U.S. Navy added to its river fleet and which was permanently lost to Dozier when it struck a snag near Cape Girardeau and sank two months after being seized. Mason managed to avoid arrest, but not the wrath of his father-in-law, who refused to have any further business dealings with him.

Since becoming part owner of the *Sultana*, Mason had evidently incurred more financial problems, for by early April 1865 he had sold most of his share in the boat, reducing his interest from three-eighths to one-sixteenth.

On April 9, 1865, the day that General Lee surrendered the Army of Northern Virginia to General Grant at Appomattox Court House, the *Sul-*

tana, with a full load of passengers aboard, docked at St. Louis, having just ended its voyage from New Orleans. Three days later it turned around in the river and began its return trip to the Crescent City. It arrived at Cairo about one o'clock the next morning, April 13. While it still lay docked there, on the evening of April 14, John Wilkes Booth shot President Lincoln in Ford's Theater in Washington, D.C. The following morning, the same morning that the president died of his wound, the *Sultana* shoved off from the wharf at Cairo and resumed its downriver run.

The *Sultana* arrived at Memphis on the morning of April 16 and departed that same morning, on its way to Vicksburg. Near Vicksburg, at Camp Fisk, a large number of Union soldiers who had been recently released from Confederate prisoner-of-war camps were waiting to be shipped to Camp Chase, Ohio, near Columbus, and Jefferson Barracks, Missouri, near St. Louis, to be mustered out of the Army before being allowed to return to their homes. The U.S. Army would pay five dollars for each enlisted man and ten dollars for each officer to be transported from Vicksburg to those two destinations. Learning about the released POWs, Mason determined to get a boatload of those veterans as passengers on the *Sultana*. They represented a potential boon to him—perhaps a partial solution to his money problems. Moreover, he figured the *Sultana* was entitled to a load of those passengers because it was one of the steamers of the Merchants' and People's Line, an organization of independently owned steamboats that had been formed two months earlier and that held U.S. government contracts to transport freight and troops.

No sooner had the *Sultana* docked at Vicksburg than Mason was talking to Lieutenant Colonel Reuben B. Hatch, chief quartermaster for the U.S. Army's Department of Mississippi, about picking up his share of military passengers when the *Sultana* made its return trip from New Orleans. Hatch assured him the *Sultana* would get some of the troops.

The *Sultana* arrived at New Orleans on April 19. At about ten o'clock in the morning on Friday, April 21, with seventy-five passengers aboard, as well as a hundred live hogs and sixty mules and horses and other cargo, it backed away from the Gravier Street wharf and began its voyage upstream. The river, running high, was swollen with the waters of the usual spring run-off from melting snow and ice in the northern reaches of the Mississippi and its tributaries. In many places it had surged out of its banks and was streaming southward wide, fast and cold.

With ten hours to go before the *Sultana* was scheduled to reach Vicksburg, its chief engineer, Nathan Wintringer, noticed steam escaping from a crack along a seam in one of the boilers of the larboard engine. He decided the *Sultana* could continue on to Vicksburg, but at a slower speed, and that repairs would have to be made at Vicksburg, where the boat had had to undergo boiler repairs once before.

It reached Vicksburg in the late afternoon on April 23, and Wintringer hustled ashore to find R.G. Taylor, a Vicksburg machinist, and have him assay the boiler problem. Taylor obligingly went aboad and upon inspection of the boilers, found a bulge in the seam of the middle larboard boiler. He asked Wintringer why the boiler hadn't been repaired before the boat left New Orleans (an indication perhaps of the danger he believed the bulge presented). Wintringer replied that the boiler was not leaking when the vessel was at New Orleans.

Captain Mason then entered the conversation and instructed Taylor to fix the bulging seam as quickly as he could. Taylor responded that to do the job right, he would have to replace two of the boiler's iron sheets and, apparently encountering some resistance, said that if he were not allowed to make the repairs he felt were necessary, he would make no repairs at all. He then turned and strode off the boat.

Right behind him went Wintringer and Mason, explaining their concern over the amount of time that it would take to replace the two sheets and asking Taylor instead to do the best he could within reasonable time constraints. Mason promised Taylor that when the *Sultana* reached St. Louis, he would make the more extensive repairs that Taylor said were needed. But for now, he urged Taylor, for the sake of time, to simply rivet a patch onto the boiler to take care of the leak. Charmer that he was, he at last persuaded Taylor. A metal patch twenty-six by eleven inches and a quarter-inch thick would be applied to the faulty boiler. The patch would be thinner than the iron sheets that formed the boiler, which were one-third of an inch thick. Taylor's idea was to smooth out the bulge first, then apply the patch. He was talked out of that and instead was asked to attach the patch over the bulge.

Time was the big consideration for Mason. He knew that those released prisoners of war were waiting for transportation, and he was apparently worried that he might lose his share of them if the *Sultana* were delayed too long. Major General Napoleon J.T. Dana, commander of the Army's Department of Mississippi, had made known that he wanted those veterans returned home as soon as possible, and two steamers, the *Henry Ames* and the *Olive Branch*, had already embarked a load of them and left Vicksburg earlier that day. The *Sultana* had to be repaired quickly, lest all the troops depart by the time lengthy repairs were done. Those potential passengers had become extremely important to Captain Mason.

Mason wasn't the only one whose mind was on something other than safety and the welfare of the soldiers who were to become passengers. The

U.S. Army officers in charge of the arrangements for transporting the men were showing their own lack of good judgment. Despite the availability of other steamers, including the *Lady Gay*, another Merchants' and People's Line steamer, which had a greater capacity than the *Sultana* but had been refused any of the soldiers, and the *Pauline Carroll*, which sat virtually empty at the Vicksburg waterfront, all of the estimated 2,400 released soldiers were crammed onto the *Sultana*. First came 398 soldiers who had just been released from the nearby Army hospital. Then came the first trainload, about 570 men. Then came a second trainload, about 400 men. Then came a third trainload, about 800 men. And all the while, despite gentle protests that the *Sultana* was being overloaded and that the *Pauline Carroll* was standing there empty and ready to go, the officers in charge, Army captain George Williams, the U.S. officer in charge of the U.S.-Confederate exchange of prisoners, and Army captain Frederic Speed, temporarily in charge of the released Union prisoners, refused to put any of the men aboard any boat other than the *Sultana*.

Williams claimed that the *Pauline Carroll* had offered a bribe of twenty cents per man to get the former POWs as passengers and for that reason it could not have a single one of them. Besides, he said, "I have been on board [the *Sultana*]. There is plenty of room, and they can all go comfortably."

The loading continued on into the darkness of April 24, and when all were aboard, the *Sultana* was literally loaded to the gunwales, from stem to stern. A long row of double-decker cots was set up through the center of the saloon to provide beds for the officers. The enlisted men, the great mass of the troops, had at first been assigned sections of the boat according to their units, the men of the Ohio regiments going to the hurricane deck and filling it, the men of the Indiana regiments going to the boiler deck, encircling the cabin, but as the day had worn on and the troops waiting on the wharf mingled with one another, order and control were lost, the rolls becoming confused and the men allowed to claim space aboard the boat wherever they could find it, men from Michigan, Tennessee, Kentucky and Virginia, most in their early twenties, many looking like skeletons from their months of starvation at the Cahaba and Andersonville prison camps, all eager to get home after having survived the ordeals of the war and the camps.

They filled the main deck, they crowded the hurricane deck, they climbed and took over the roof of the texas deck, the mass of their bodies so heavy that William Rowberry, the *Sultana*'s first mate, directed a work party of deckhands to wedge stanchions between the boiler deck and the hurricane deck to prevent the sagging upper decks from collapsing.

To sleep, the troops would have to lie down anywhere they could, on the steps of companionways, beside the engine room, most on the open decks, stretched out like pieces of cordwood beside one another. In the absence of restrooms, they would contort themselves over or under the boat's rails to relieve themselves.

William Butler, a cotton merchant from Springfield, Illinois, who from the deck of the *Pauline Carroll* watched the third trainload of troops board the *Sultana*, reported the troops' reaction: "When about one third of the last party that came in had got on board, they made a stop, and the remainder swore they would not go on board. They said they were not going to be packed on the boat like damned hogs, that there was no room for them to lie down, or a place to attend to the calls of nature. There was much indignation felt among them, and among others who went about the boats. Some person on the wharfboat, an officer I presume, ordered them to move forward and they went on board."[8]

About nine o'clock in the evening on April 24 the *Sultana* at last backed away from the Vicksburg wharf and resumed its northward voyage, the men making the best they could of their miserable situation, which must have seemed but an extension of the horrors they had suffered in the prisoner-of-war camps. Through two nights and two days they endured as the *Sultana* steamed northward toward home.

About one o'clock in the morning of Thursday, April 27, 1865, having loaded aboard enough coal to get it to Cairo, the *Sultana* pulled away from the coaling station at Memphis and again headed upstream, into a particularly dark night and misting rain. All was quiet on the decks, in the saloon and in the staterooms, almost all of the passengers and most of the crew deep in slumber as the lights of Memphis receded and disappeared. About seven miles above Memphis the *Sultana* approached a group of marshy islands called Paddy's Hen and Chicks. Ordinarily the river at this spot was three miles wide, but now, swollen with floodwaters, it was three times that width, its waters covering both banks.

At one-forty-five the *Sultana* passed one of the Chicks, island No. 41, near the Arkansas side of the river. After that came Fogelman's Landing. It was now two o'clock.

Suddenly the *Sultana* erupted in a massive explosion, as if struck by a thunderous earthquake, shattering the vessel, blasting a chasm in the superstructure above the boilers, spraying deadly scalding steam over passengers, blowing parts of the upper decks to fragments and shooting them and passengers high into the night sky, toppling the two chimneys, hurling most of the pilothouse from its perch atop the texas, scattering debris and human bodies, many broken and dead, others still alive but gravely injured, into the cold, dark, surging river. Burning coals, blasted from the vessel's fireboxes,

The ill-fated steamer *Sultana*, victim of the worst maritime disaster in U.S. history, photographed at Helena, Arkansas, on April 26, 1865. Early the next morning, as it steamed northward from Memphis, crammed with Union soldiers recently released from Confederate prisoner-of-war camps, it exploded and burned, taking more lives than did the *Titanic* (Library of Congress).

rose like fireworks into the air and rained back down on the wooden decks and superstructure, setting them alight. Within minutes the shattered *Sultana* was engulfed in flames.

From everywhere came the cries, shouts, shrieks and panicked screams of desperate passengers, scalded, burned or critically injured, many trapped beneath wreckage, unable to save themselves or be saved. Soon many hundreds faced the horror of being burned alive or escaping the flames by jumping into the dark and deadly river. Nearly all who were able chose the river. The badly injured and the horribly scalded begged their fellow passengers to lift them and throw them overboard, which some did. The river within yards of the boat quickly became a seething mass of struggling humanity, individuals bobbing so tight together it became impossible for more escaping passengers to leap into the water without landing first on those who had jumped before them. Those unable to swim grasped at anything, anyone within reach, and many went down to their deaths in the frantic grip of each other.

The exact total number of the *Sultana*'s victims remains uncertain, because the number of military passengers aboard is in dispute, as is the number of survivors. Many — perhaps three hundred or more — who were admitted to hospitals and were initially counted as survivors died of their burns or injuries within days after being hospitalized. The estimates of the total number of victims, military and civilian, range from 1,238 (the estimate of Brigadier General William Hoffman, who conducted an inquiry immediately following the disaster) to 1,547 (the estimate made by the U.S. Customs Department at Memphis) to 1,800 or more (calculated from the varied estimates of the number of survivors, which ranged from fewer than 500 to around 800).[9] The likelihood is that the higher estimates of victims are more accurate.

The boat itself, "a massive ball of fire," as an eyewitness on shore described it, drifted downriver and became lodged against one of the Chick islands near Fogelman's Landing, just above Mound City, Arkansas. There it sank, its charred remains eventually buried beneath the river's sediment. In the years since, the river, as if abhorring any reminder of the disaster, shifted its course three miles to the east of Mound City, and the *Sultana*'s grave is now unknown, lying somewhere in an Arkansas farm field.

The cause of the explosion is likewise unknown. An investigation, ordered by Major General Cadwallader C. Washburn, commander of the U.S. Army's District of West Tennessee, yielded an official report, made public on May 2, 1865, that declared the cause was insufficient water in the boilers. General Dana, commander of the Army's Department of Mississippi, the man who had ordered the released prisoners sent home as soon as possible, also held an investigation, which, like Washburn's, had more to do with the overloading of military passengers aboard the *Sultana* than it did with the circumstances of the explosion. On April 30, the U.S. secretary of war, Edwin Stanton, ordered the commissary general of prisoners, Brigadier General William Hoffman, to conduct an additional investigation. Hoffman's report concluded that there was not enough evidence to establish the cause of the explosion, but that the evidence did suggest that insufficient water in the boilers was the cause.

And there, insofar as the Army was concerned, the matter rested. Claims that Confederate saboteurs had placed explosives aboard the *Sultana* were apparently never seriously considered by the investigators. There is no mention of such in the testimony given to the investigators, and nothing in the investigators' inquiries indicates sabotage was even suspected.

The relevant civilian authority, J.J. Witzig, the supervising inspector of steamboats for the area that included St. Louis, faulted the patch riveted onto the leaking boiler and put the blame for allowing it on the *Sultana*'s chief engineer, Nathan Wintringer. Witzig, however, conceded that the tubular boilers were part of the problem.

In the end, no one was held accountable for the *Sultana* disaster. The guilt was eventually assigned to the design of the boilers. After the *Walter R. Carter*, equipped with tubular boilers like the *Sultana*'s, blew up later in 1865, killing eighteen people, and the *Missouri*, similarly equipped, exploded the next year, killing seven, the offending boilers were removed from all steamers traveling south of Cairo.

Part Four. The Outcome

• 12 •

On to Cairo

Before reaching Vicksburg late Friday afternoon, Tom Leathers was pleased to see that the *Natchez*, fiercely striving to catch up, was gaining, though slightly, on the *Robert E. Lee*. Checking his watch, he estimated that, except for the head start the *Lee* had got at New Orleans, the *Natchez* was running only about eight minutes behind, a lead that was not at all insurmountable. But he had lost more time at the Vicksburg wharf, putting off passengers and their baggage and taking on fuel. That done, he had steamed off again, renewing his hot pursuit, but with more time to make up.

In the early evening, just above the mouth of the Yazoo River, as the *Natchez* approached Buckhorn Landing at Milliken's Bend, Leathers's chief engineer, Andy Pauley, discovered that the pump that drew river water into the boat's boilers had suddenly quit and could not be restarted. He diagnosed the problem as a broken valve. There was nothing to do but head the *Natchez* into shore, tie up to a tree on the riverbank, shut down the engines, remove the valve and fix it. Pauley made the repairs, but consumed thirty-four precious minutes in the process. Then he started up the engines again, and the *Natchez* carefully backed out into midstream and resumed its urgent run, its captain fuming in frustration.

Then came a new misfortune. In the darkness of Friday night the speeding vessel, despite the reputation of its pilots, lost the channel and ran into shallow water near the shore off island No. 93 and grounded on the river bottom. Only after more anxious minutes were the pilot and the engineers able to dislodge the boat by reversing its huge paddle wheels. It then returned to the safety of the main channel, where it straightened up in deeper water and re-entered the race.

Under way again, the *Natchez* passed the steamer *Frank Pargoud*, which was headed downstream. Leathers had heard reports that Cannon had arranged to have the *Robert E. Lee* refueled from the speedy *Pargoud* rather than tying the *Lee* up to burdensome barges to take on fuel. The passing of the *Pargoud*, whose crew failed to answer the *Natchez*'s hail, a sign that something was not quite right, silently told the enraged Leathers that the reports were true.

The *Frank Pargoud* maneuver was indeed Cannon's latest trick. The *Pargoud* was owned by John W. Tobin, Cannon's long-time and wealthy friend who was aboard the *Lee*, giving Cannon moral support and expert advice as well as enjoying the history-making ride. The *Pargoud*'s usual run was between New Orleans and Greenville, carrying passengers and freight. But on the night of Friday, July 1, 1870, it had no passengers, no lights beaming from its stateroom windows, and the only freight it bore was one hundred cords of knotty pine wood, oozing thick, sticky sap that would make it burn bright and hot, just the fuel needed for a racing steamboat. By prior arrangement, Tobin's boat had stood idling in mid-stream just below Greenville, waiting for the *Robert E. Lee*. When the *Lee* had appeared, between two and three o'clock in the morning, the *Pargoud*, its bow pointed upstream, had steamed up alongside the swiftly moving *Lee* as it passed. Lines had gone out, lashing the boats together, and the two steamers had raced side by side while brave and hardy roustabouts walked gangplanks laid between the boats, carrying armloads of firewood from the main deck of the *Pargoud* to the main deck of the *Lee*.

While the two boats were tied together, with planks between them, Governor Warmoth and Doctor Smyth, both with other matters doubtlessly on their minds, had seized the opportunity to disembark from the *Lee* and had made their way across the gangplank to the *Frank Pargoud* to return to New Orleans. When the transfer of wood and the two passengers was completed, the *Frank Pargoud* had cast off the lines that bound it to the *Robert E. Lee*, and its captain had let his boat fall behind the *Lee*, then had made a big U-turn in the river and headed back downstream, a dangerous night's work efficiently done. Meanwhile the refueled *Robert E. Lee* continued on through the darkness, rushing toward Helena and Memphis.

The *St. Louis Republican* reporter aboard the *Natchez* cried foul, as did fellow passengers who had bet on the *Natchez* and learned about the *Frank Pargoud* incident by way of the public announcement Captain Leathers made to those still awake at that late hour. They protested that Cannon and the *Lee* had disqualified themselves by using the *Frank Pargoud* as a sort of power booster while it took on its load of fuel. "The *Lee* won all her bets up to the time when the *Pargoud* improperly and unfairly aided her by making use of her own propelling power while transferring a heavy lot of pine fuel," the reporter wrote. "The propelling power being thus divided, from another boat, loses the race for the *Lee* and all bets, notwithstanding she was in the lead. Hurrah for the *Natchez!*"[1]

Having become a *Natchez* partisan, the reporter declared his confidence in the boat's ultimate victory and the righteousness of its cause. "Everything goes lovely just now," he wrote, "and the goose hangs a trifle high; but Capt. Leathers has a fearfully long reach, and aside from the question of the bets (which the *Lee* has forfeited) the *Natchez* has a good show to make the best time. We are making a fair, open business trip, although not attempting to do much business. But we are not making a run for a race, but to try and see what can be done in the way of fast work on a regular trip."[2] When he wrote that, the *Robert E. Lee* was twelve miles ahead.

Unable to resist the chance to make a buck, Leathers slowed down at Greenville and pulled up to the wharf to take on passengers, only to discover there were no passengers waiting to come aboard. He quickly hauled his lines back in and left, after losing another ten minutes. While at the wharf, though, he had learned from someone on shore that the *Lee* was an hour ahead of him.

About ten o'clock Friday morning the *Natchez* reached the mouth of the Arkansas River, passing the site of the washed-away town of Napoleon. At White River, at eight minutes past ten, the *Natchez* slowed again, this time to tie up to and tow a barge from which it took on three hundred boxes of coal and received the somewhat good news that the *Robert E. Lee* was fifty minutes ahead. The *Natchez*, now two-thirds the way between Greenville and Helena, had gained ten minutes since leaving Greenville.

When the *Natchez* was in sight of Helena, the hopeful reporter aboard wrote: "The *Natchez* will undoubtedly set a mark that will be the goal of other boats for years to come. If we had put her through without landing, taking our fuel from steamers with full head on, and for the sole purpose of racing, we could have made Helena at least an hour ago, which is the opinion of every man on board. Helena is now in sight. We will not stop, but I will send this ashore by a skiff, if possible."[3]

At Hardin Point the *Natchez* received a warm welcome from the steamer *Mollie Able*, which stopped and swung sideways in the river, so that its bow pointed toward the *Natchez*, and it saluted Leathers and his crew with a blast from its whistle. The *Natchez* signaled its acknowledgment with a blast from its own whistle.

Continuing to keep track of the race was the New Orleans *Picayune*, which on page one on Saturday printed a dispatch telegraphed to it when the two steamers had passed Helena:

> HELENA, ARK., July 2 — The Lee passed here at 4:30 [P.M.], the Natchez at 5:24. Lee 54 minutes ahead. All her window blinds were down and some plank off her wheelhouse, and she seemed to be driving through the water. Neither boat landed here. A party went out to the Natchez in a small steamer. On their

return reported that the *Natchez* claims to have broken her pump and laid up for 30 minutes last night.

The *Lee's* time from New Orleans to this place is 47 hours 36 minutes — the fastest time on record. The *Natchez* says her time to this place is one hour and a half faster than her last trip. The *Lee* cheered with her whistle when she passed, and was answered by two steamers lying at the wharf and the multitude of people on the shore. Those who think the *Natchez* laid up 30 minutes last night are still betting on her.[4]

The *Natchez* was showing how fast it could run, breaking all of its previous speed records so far. But it was still trailing the *Robert E. Lee*.

At Memphis the excitement and the crowd waiting at the river's edge to see the racers were equally huge. The crowd's anxiety was heightened by a breakdown in the telegraph line below Memphis, the eager spectators not knowing where the boats were or when to expect them at Memphis. A news reporter in Memphis described the situation:

All coming from the [Memphis telegraph] office were eagerly questioned, but no news could be gained, as the operator at Helena reported the boats not in sight at 3:45 [P.M.]. With every moment's delay the excitement increased. At one time it seemed probable that the curiosity of the populace would not be gratified till the boats arrived at this place [Memphis], for the telegraph line between Helena and Madison was blown down by a sudden storm at 4 o'clock this afternoon. When this news was bulletined, a verbal cry of disappointment arose, and murmurs that some ruse was being practiced found many believers. But fortunately the telegraph operators were equal to the emergency. Repair men were immediately sent out from different stations, and by 7 P.M. the line was again in order, and the news came that the Lee had passed Helena at 4:30 and the Natchez at 5:24....

The curiosity of the people to see the boats as they pass is intense. Many have been on the bluff [on the river] all the afternoon, and since dark the crowd has increased till the whole bluff is now covered, and still people are coming in from all parts of the city.

Great preparations are being made for the reception of the boats. Tar barrels are placed ready to be fired as they approach, and a battery of artillery is in position ready to thunder forth a salute in honor of the victor. All seem wild with anticipation. Men, women and children are striving for favorable positions to witness the race, and all seem animated with an intense desire to gain a good look at the boats as they pass the city.[5]

Among the crowd, absorbed with the race, rumors, all false, were rife. One had it that the *Robert E. Lee* had broken down and was unable to continue the race. Another was that the larboard wheel housing of the *Natchez* had been blown off and that the *Lee* was towing the *Natchez*. Still another claimed that the *Natchez* was gaining on the *Lee* and that Captain Cannon had despaired of winning the race. Other rumors asserted that the recent telegraphic message reporting that the *Lee* had passed Helena fifty-four minutes ahead of the *Natchez* was a fraud perpetrated by friends of the *Lee* who had tampered with the telegraph wire.

The *Lee* had been expected to arrive about nine P.M.—despite Captain Cannon's promising only to make Memphis by eleven—and the steamer *Connecticut* had left the Memphis wharf earlier that evening carrying a capacity load of passengers, at two dollars a head, all intent on seeing the two boats steam by. At ten P.M. the *Connecticut* was still out in the river, its passengers waiting and wondering where the racing boats were. The reporter covering the event at the Memphis waterfront seemed to grow as anxious as the other spectators:

> 10 P.M. The boats are not in sight yet. The crowd on the bluff in front of the river is immense. Nothing like it has been seen in this city for very many years. Seats have been impromptued for the ladies, and the whole front of the city looks like one vast amphitheatre, and the utmost interest is manifested on all sides.
>
> Betting is very heavy, with the chances slightly in favor of the Lee. But experienced steamboatmen claim that the Natchez can more than make up her lost hour above this point, and great confidence is still manifested by the backers of the natty champion.
>
> Tugs have been stationed in the river with barges of coal ever since 8 o'clock, but up to this hour nothing has been heard of either boat.
>
> Old men who have not been to the levee for years came out on this occasion, to gaze upon the magnificent spectacle. Carriages, hacks, buggies and every specie of vehicle crowded the levee. Bonfires are prepared and will be fired as soon as the boats come in sight.[6]

The people of Memphis, the city itself, were planning a spectacular welcome for the race's leader, which at last report was the *Robert E. Lee*. Not only the cheering crowds, the artillery salutes and the bonfires, but a breathtaking fireworks display were awaiting the *Lee*, the fireworks to be set off the moment the *Lee* was spotted clattering and splashing up the river, its cabin lights glowing from a distance, gleaming through the night's blackness.

Then suddenly at twenty minutes past ten shouts of "Here she comes!" burst from the crowd as the lights of an approaching steamer, still far off, appeared and all eyes turned downriver. It was the signal that the celebrants had been waiting for. The barrels filled with tar were set ablaze, their flames quickly leaping into the night, firecrackers popped, cannons thundered, and barrages of skyrockets shot up, brightening the dark sky with brilliant color—an altogether dazzling winner's welcome to the city of Memphis. But as the approaching steamer came closer, the cheering spectators along the riverfront could see it wasn't the *Robert E. Lee*. It wasn't the *Natchez* either.

The *Thompson Dean* had left New Orleans a few days before the *Lee* and the *Natchez*, making its regular, comparatively unheralded run between New Orleans and Memphis, and here it came, as if crashing the party, its crew and

Riverfront at Memphis in 1862. The spectacular welcome that the people of Memphis planned for whichever boat was in the lead when the racers reached Memphis was set off prematurely when in the darkness the excited crowd mistook the steamer *Thompson Dean* for the *Robert E. Lee* (Library of Congress).

passengers doubtlessly embarrassed by the mistaken fuss it had set off, its captain sounding its whistle in sheepish response as the boat slowed and glided up to the wharf.

The cannons then were reloaded, and a search for more fireworks was begun. The premature display was not entirely wasted, however. So brilliantly arresting were the fireworks that those on board the *Robert E. Lee* could see them in the sky as the *Lee* came by Presidents Island, in the great bend of the river just below Memphis.

A half hour later, at 11 P.M., the reporter in Memphis, recording every new, significant occurrence at the waterfront — although not bothering to mention the crowd's awkward mistake — told his readers: "A bright light is just coming into view around President's Island, six miles from the city. It is believed to be the *Lee*."

This time it *was* the *Lee*. Quickly the reporter added to his dispatch: "11:04 — The *Lee* has just arrived and is taking coal barges in tow. Enthusiasm is immense. The crowd is cheering, cannons firing and bonfires blazing."[7]

Two more terse adds to his dispatch told the rest of the story:

> 11:10 — The Lee has just left. Such an ovation has never been given to any boat before. The people are wild and bets are freely offered that she will beat the Natchez to Cairo one hour and fifteen minutes.
>
> 12:13 — The Natchez has just passed. The crowd is fast dispersing and a day of great excitement is over.[8]

The *Natchez* had stopped long enough to put fourteen passengers ashore, then had slipped back out into the river and taken on coal from a pair of barges that had been waiting for it, losing another seventeen minutes in so doing. By the time both boats had passed Memphis, the *Lee*'s enthusiasts, seeing its big lead, were so ecstatic over the race's progress that they were offering odds of ten to one that the *Lee* would reach St. Louis first.[9]

The course northward from Memphis took the racing steamers past and through the group of islands called Paddy's Hen and Chickens, site of the disastrous explosion and fire aboard the *Sultana* five years earlier, and twisted through the narrow, shallow channels in the darkness, which the *Natchez*'s adherents must have thought favored their boat, with its shallower draft. However, as the *Natchez*, running at racing speed again, steamed into the night, it ran into more bad luck, plowing into shoals again, off island No. 41 (the forty-first island in the river, counting downstream from Cairo), scraping its already bruised hull and once more forcing its pilot to back up repeatedly and churn loose from the river's muddy bottom. The result was more lost time.

Through the labyrinth of islands the *Natchez* continued, threading its way past island No. 38, where it executed a sharp turn, then past Devil's Bend and No. 37 and at last emerging onto a straight stretch that let it resume full speed.

Around five-thirty on Sunday morning, July 3, it drew abreast of Fort Pillow, then, just above the fort, it reached Plum Point, where on the morning of May 10, 1862, the Confederacy's meager collection of civilian steamboats converted into warships to defend the Mississippi — the *General Bragg* (formerly the *Mexico*), *General Beauregard* (formerly the *Ocean*), *General Sterling Price* (formerly the *Laurent Millaudon*), *General Sumter* (formerly the *Junius Beebe*), *General Lovell* (formerly the *Hercules*), *General Jeff Thompson* (former name unknown), *General Van Dorn* (former name unknown) and the *Little Rebel* (formerly the *R & J Watson*) — won their lone victory over a Union fleet of ironclad gunboats, then had fallen back to Memphis to face a devastating defeat four weeks later.

On the *Natchez* raced, through another set of islands scattered in a narrow stretch of river, then back onto a wide straightaway, past Needham's Island, which once had been a salient projecting into the river but now, cut off from the Arkansas shore by erosion, an island in the stream, then on beyond islands 21 and 20, then past Nos. 19 and 18, opposite the Missouri state line on the western shore.

Near island No. 14, shortly before Sunday noon, the *Natchez* came upon the steamer *Belle of Memphis*, approaching from upriver. While it was still a distance away, the *Belle of Memphis* courteously stopped her engines and paddle wheels to give the *Natchez* calm water as it steamed by, then saluted the racer with a whistle blast, which was returned by the *Natchez*.

At No. 14 Captain Leathers again consulted his watch. Two days, eighteen hours and sixteen minutes had elapsed since he had sped the *Natchez* past St. Mary's Market, making the present run two hours and forty-two minutes better than the boat's previously best time — but still running behind the *Robert E. Lee*.

The *Lee* had passed No. 14 almost exactly an hour earlier and was about fifteen miles in front of the *Natchez*. Passing the Tennessee-Kentucky border on the eastern shore, the *Lee* steamed on toward New Madrid, Missouri, on the west bank, and reached it around one P.M. Then came a straight, twelve-mile stretch that ended at island No. 10, situated in an S-shaped curve above New Madrid. Island 10 was another memorable Civil War site, where the guns of a Confederate fortification had checked the Union Navy's advance down the Mississippi until the U.S. gunboat *Carondelet* bravely ran past it. The *Carondelet* was captained by Commander Henry Walke, a daring and imaginative steamboat naval officer who insulated his vessel with cordwood, hawsers and chain and tied a coal barge loaded with hay to the gunboat's larboard side to ward off shot as the *Carondelet* steamed defiantly downriver to the right of the island on the night of April 4, 1862. On April 6 the *Carondelet*, from its position below No. 10, turned its guns on the island's batteries and, joined by the gunboat *Pittsburg*, knocked out several of them. The Confederates, seeing their position was no longer tenable, withdrew from the island and abandoned their fortification, opening the way for the U.S. gunboats' drive on Memphis.

At island No. 8, roughly a thousand miles from New Orleans, some of Cannon's friends aboard the *Robert E. Lee* noted its elapsed time since passing St. Mary's Market as two days, twenty-one hours and seventeen minutes and computed the *Lee*'s average speed at more than fourteen miles an hour since the start of the race.

As the *Lee* passed No. 8, the town of Hickman, Kentucky, came in sight on the east side of the river. From Hickman a *St. Louis Republican* reporter filed the latest dispatch on the race:

> The Lee passed the wharf here at 3:41 P.M. [Sunday], railroad time, hurrying with unparalleled velocity. Clouds of spray were right and left by her hurrying bow, and the air was blackened with dense columns of smoke that issued from the heated chimneys. When the Lee appeared, the smoke from the Natchez was hardly visible twenty miles or more below. The whole population of the place — men, women and children — had been anxiously awaiting the arrival of the racers, and never before was there such a multitude upon the river banks as on this bright Sabbath afternoon.... As the Lee came up, the multitude greeted her with

12 • On to Cairo

cheers loud and long. Two cannons, placed upon the wharf, were fired at short intervals, and the excitement of the multitude that thronged the river shore for a mile or more was boundless....[10]

The reporter at Hickman also gave news of the *Natchez*, as well as his opinion of its chances: "The Natchez passed up at 4:45. There was great cheering, and the guns were again fired. The Natchez was gaining steadily on the Lee as she passed, but she is hopelessly beaten."[11]

From Columbus, Kentucky, the site of another strategic fortification lost by Confederate troops guarding the Mississippi, the *St. Louis Republican*'s reporter telegraphed:

The patience of the good people of this city and vicinity, who have been congregated along the river front nearly all day, has just been rewarded. The Lee passed at 4:41, apparently in splendid condition, driving through the water like some magic marine monster. She was munificently cheered. The Natchez is not heard from.[12]

The reporter later telegraphed to say: "The *Natchez* passed here at 5:51 ½."[13]

Above Columbus came Lucas Bend, where island Nos. 4, 3 and 2 clustered, tightening the channel of the river. Then came island No. 1, and beyond it stood Cairo behind its protective levees.

At Cairo, at the juncture of the Ohio and Mississippi, crowds of onlookers had poured into town, some traveling considerable distances by train, to view the passing of the steamers, and by early Saturday evening many of them had begun to cluster and camp along the levee, making sure they would not miss the big event, uncertain about when exactly the racers would arrive. Both boats would have to refuel at Cairo, and the spectators figured they would get a good, long look at them at that coaling stop. A dispatch from Cairo published in the *Picayune* described the scene:

The levee here swarms with people come from far and near to witness this historical race. Not only has the whole population of Cairo, men, women, children, the infirm and the aged, without regard to race, color, sex or previous condition of servitude, turned out, but strangers from St. Louis, from Cincinnati, Louisville and the Ohio River towns, and from the railroad towns in Illinois, are here too, standing at the same time on the banks of the Mississippi....[14]

As the *Robert E. Lee* came up to island No. 1 Cannon and his passengers could see the steamer *Idlewild* standing in the river. The *Idlewild*, its run ordinarily being between Evansville and Cairo, had come down the Ohio, stopped at Cairo, then had swung southward down the Mississippi and stopped at island No. 1, about twelve river miles below Cairo, to wait for the *Lee*. Aboard the *Idlewild* were some three hundred passengers, including a large group of excursionists who sought a close-up view of the racers, a number of other passengers who wanted to board the *Lee* and ride it to St. Louis on the final leg of the course, and two pilots, Enoch King and Jesse Jameson, who were especially knowledgeable of the vagaries of the river between Cairo and St. Louis, a stretch not so familiar to the *Lee*'s regular pilots, Wes Conner, James Pell and George Clayton, who were probably thankful to be relieved from their tasks during this critical last leg of the race.

Ever thinking ahead, Captain Cannon had made arrangements for the *Idlewild* to meet the *Lee* and bring him the pilots as well as to take from the *Lee* the passengers from New Orleans who had bought passage to Louisville and other stops along the Ohio. Those passengers would be transferred to the *Idlewild* in mid-stream, in the same way the *Lee* had taken on firewood from the *Frank Pargoud*.

Once the *Lee* was sighted, around five o'clock, Captain Gus Fowler of the *Idlewild* restarted his engines and as the *Lee* drew near, the *Idlewild* steamed upstream, its captain hoping to keep up with the *Lee* and, as one eyewitness believed, match its speed against the *Lee*'s. That idea was banished in a few minutes, assistant engineer John Wiest related, the *Lee* having to slow down to allow the *Idlewild* to stay abreast of it. The *Idlewild* performed the same act as the *Pargoud* had done, Wiest reported, casting off as soon as the passengers and baggage had been transferred, while the *Lee* resumed full speed, passing Cairo.[15]

Having reached Cairo at six P.M., the *Robert E. Lee* had made it there from New Orleans in three days and one hour, beating the previous best time, made by the *A.L. Shotwell*, by two hours and forty minutes, thereby laying claim to another set of horns.

Cannon had also made arrangements for refueling at Cairo. Waiting near the Missouri side of the river was the steam tugboat *Montauk* with four barges loaded with coal, two for the *Robert E. Lee* and two for the *Natchez*. Perhaps because Cannon's coal order had come to the supplier earlier than that of Captain Leathers — whose telegram ordering the coal and the arrangements was not received until Sunday, the day the coal was needed — the two barges of coal intended for the *Lee* were positioned toward midstream in the river for a relatively easy pick-up, and the two intended for the *Natchez* were positioned toward the Missouri shore, in shallow water that could pose a danger for the *Natchez*. The *Lee* slowed down to pick up the waiting barges. One was tied to the *Lee*'s larboard side, and the other to its starboard as it continued upstream, its crewmen manhandling 1,500 bushels of coal aboard the *Lee* in the process.

Also coming aboard the *Lee* were a number of newspaper reporters, who boarded the vessel on the fly. One, apparently desperate to share in some of the race's drama, related the experience to his editor: "Owing to the fact that neither the *Lee* nor the *Natchez* landed here [Cairo] your correspondent was so engaged in making arrangements to get on board and off again, that no regular report can be sent to you. Dropping on board the *Lee* in mid-river as though shot from a cannon, I managed to obtain the following items during the two-mile ride...."[16] He then went on to succinctly relate high points of the *Lee*'s voyage, information that had previously been published.

Another reporter, evidently no fan of John Cannon, offered his commentary on the race as viewed from the Cairo waterfront:

> It is precisely 6:10 P.M. and the steamer Lee has just passed a point opposite Cairo ... bound for St. Louis. She has no time to pay her respects to Cairo in respect for the prolonged shouts upon shouts that the multitude threw across the water toward her as she passed.
>
> The avenger is on her track, and though she is ahead of the White's time and ahead of the Natchez's time, and ahead of everything in the world except herself, she has no leisure for courtesies and compliments.
>
> The race is considered virtually ended, without an accident between here and St. Louis to delay the Lee till the Natchez can overhaul her.
>
> But she is not content with simple victory. She has spent much labor and sacrificed much money to prepare for this race and is determined to set her peg where it will not be pulled up soon. The J.M. White's time remained untouched for twenty-six years. Capt. Cannon wants his boat's time to remain untouched to the end of the present century.[17]

At 7:08 P.M. the *Natchez*, having slowed to pick up its six hundred boxes of coal on the run, as the *Robert E. Lee* had done, without mishap, passed Cairo, now running about twenty miles behind the *Lee*.

Both boats were entering the final and most crucial leg of the race, from Cairo to St. Louis, with night quickly enveloping the river and the water at a low stage. A steamer descending from St. Louis, the *Rubicon*, had arrived at Cairo just minutes earlier than the *Lee* and had reported that there was but eight feet of water in the main channel above Cairo. To make matters worse, the Mississippi above Cairo was an obstacle course of islands, rocks, sandbars and narrow, often perilously shallow channels. Because of those dangers, some steamer captains routinely refused to take their boats on the run between Cairo and St. Louis after dark.

But Captain Cannon, relentlessly racing northward, was undaunted. And Captain Leathers had no choice but to follow.

• 13 •

The Fog

Not long after it had pulled away from Cairo, the *Natchez* engaged its first obstacle. Near the Illinois shore it ran aground on a sandbar, which cost it more time as it backed and struggled to free itself, then, managing to escape, sped forward again as night fell upon the river. Minutes later a new menace, more blinding, more threatening than nightfall, came seeping silently over the dark river. Fog. In places along the river's course the gray fog was a wispy veil. In other places it was too thick to be penetrated by the eye. The riverbanks disappeared behind it.

The *Natchez* now was passing its obstacles with extreme caution, slowly hugging Elk Island, staying to starboard of it, slipping past the pair of mid-river islands called the Two Sisters. At Dog Tooth Bend the boat struck bottom again. More time lost. When it was free again, the *Natchez*'s pilot showed more caution, feeling his way along, stopping and reversing engines when he suspected a threatening sandbar, before ramming into it.

The wits among the passengers aboard the *Natchez*, having abandoned all hope of catching up to the *Lee*, began making jokes about the *Natchez*'s creeping pace. The best chance of seeing the *Lee* again, one wag remarked, was on its return trip from St. Louis. The *St. Louis Republican*'s reporter aboard the boat, at last giving in to the *Natchez*'s bleak situation and faintly critical of the *Natchez*'s pilots, wrote that "the race with the *Lee* ... was virtually ended, unless the pilots of the latter should go crazy and jump overboard, and even in such a contingency there would be but little chance for the *Natchez*."[1]

Captain Leathers, fulminating with curses and other blue language, apparently heard none of it and would not have been deterred if he had. He continued to press his pilots and other crewmen on. At another sharp twist in the river, called Hackett's Bend, the *Natchez* grounded again, with a long scrape of the hull followed by a thud that halted forward movement. The reporter from the *St. Louis Republican* observed that "it really seemed we were engaged in a sleighing expedition instead of a steamboat race."

It was now after nine P.M. The *Natchez* had left Cairo more than two hours earlier and had come but twenty-two miles. It was running out of the channel and into the river bottom about every two miles, suffering more delays each time it had to free itself. So much for the claim that the *Natchez* had the best pilots on the Mississippi.

The fog was deepening, but Leathers kept the *Natchez* inching through it, warning crewmen he had posted on the decks to keep a sharp eye out for whatever lay hidden ahead. It was clear to crewmen and passengers alike that he had no intention of stopping. He knew that the fog and low water must be hampering Cannon and the *Robert E. Lee* as well as the *Natchez* and all he had to do to stay in the race was to keep going. After all, the *Lee*, with its deeper draft, was more susceptible to grounding than was the *Natchez*.

He managed to make it past the shallows off Goose Island, then past the jagged, underwater rocks of the Grand Chain and past the landing at Thebes, Illinois, the lights of which could be seen through the hazy fog, which was just a thin cloud at that point. Above Thebes the river straightened, and the run got easier, except for the threat presented by a rocky islet that the *Natchez* must pass to larboard before continuing on into a deep channel. The *Natchez* slid safely past the islet, turned westward and sought the reassuring wharf lights of Cape Girardeau, which, it turned out, were barely visible through the worsening fog.

Above Cape Girardeau the river was again strewn with rocky hulks jutting up from below the surface, a section of the Mississippi appropriately called Devil's Country. The little islands, not numbered as the islands below Cairo were, were known by such hellish names as Devil's Island, Devil's Tea Table, Devil's Bake Oven and Devil's Backbone. After them came Dog Island and Muddy Island, only slightly less forbidding. As the *Natchez* groped toward them, the fog became nearly impossible to deal with. Leathers tried to cope, inching the *Natchez* through the shrouded chain of rocky islets and doubtlessly wondering if he might at any moment come across a stranded *Robert E. Lee*, run afoul of the ruinous rocks — and perhaps hoping that he would. Now Leathers's pilots were ready to give it up, the hazards being too great. All they could see, and then only fleetingly, was a small patch of water directly off the bow. Leathers, though, insisted that they keep going.

Somewhere around Hamburg Island Leathers consulted his watch and learned that midnight was just minutes away. Evidently tired and expecting nothing but more tension and danger ahead, he at last decided he had had enough fog for one night. With some ninety passengers aboard and his boat and crew at risk, facing the mounting danger of smashing into fog-hidden rocks, Captain Leathers gave up the idea of a continued pursuit — for the time being. Around half past twelve, he ordered his pilots to feel their way toward the Missouri shore and get close enough for the deckhands to go ashore and tie the *Natchez* to a tree, and there they would impatiently wait for the fog to lift.

Warily braving the threat of sudden grounding, pilots Frank Clayton and Mort Burnham groped through the thick fog and managed to find the riverbank without running into it. Once the vessel was secured they ordered the *Natchez*'s engines stopped and resigned themselves to an indeterminate delay.

The place where they landed turned out to be Kinney's Point, at the top of Devil's Island, close to Shepherd's Landing, the site of a woodyard run by a man named Delvory. Apparently thinking he had acquired a customer, Delvory at that late hour came walking through the fog to talk to the officers of the *Natchez*. From Delvory Captain Leathers learned that the *Robert E. Lee* had crept by Shepherd's Landing not more than twenty-five minutes earlier. In the clutching fog, the *Lee*'s lead had shrunk to just twenty-five minutes. It still could be caught, Leathers believed, if the fog would dissipate enough for the *Natchez* to get back in the race. It was now 12:35 A.M., Monday, July 4.

The *Lee*, with its St. Louis pilots at the wheel, had steamed briskly from Cairo in daylight, and its passengers had happily received the salutes of passing steamers on their way downriver, the *Nick Wall*, the *St. Joseph* and the *Olive Branch*, before the sun set behind the trees that crowded the Missouri shore. Once the sun had sunk out of sight, the air swiftly took on a damp chill. On the Missouri side of the river bonfires were lighted, for the racing steamers or the imminent Fourth of July, or both.

At Cape Girardeau, which it reached about nine-thirty, some four hours and forty-four minutes out of Cairo, the *Robert E. Lee* was greeted by more bonfires. It was there that those aboard the *Lee* noticed the wispy haze drifting over the river, dulling the brilliance of the bonfires' flames. Fog was moving in.

As the fog thickened, the *Lee* kept moving, but Captain Cannon called a conference with the St. Louis pilots, Jesse Jameson and Enoch King, and his regular pilots as well as others whose judgment he trusted. After conferring, and with the approval of Jameson and King, Cannon decided to keep moving despite the fog, which, deep into it, pilot Jameson claimed was the worst he had ever encountered in some twenty-five years on the river. "Ordinarily," assistant engineer John Wiest reported later, "the boat would have laid up to wait for it to clear away, but those St. Louis pilots were game and never said anything about quitting."[2]

What those pilots did say, though, was that they wanted the help of "the best eyes on board,"[3] which they got, to help them keep the boat from harm.

They also got some other extraordinary precautions ordered by Cannon. Cannon stationed both Jameson and King in the pilothouse and put his other pilots — Conner, Pell and Clayton — at the forward end of the texas deck, closer to the bow than were Jameson and King in the pilothouse. He then positioned three other sharp-eyed, river-wise crewmen on the main deck, at the boat's prow. And he lowered the *Lee*'s yawl into the river, manned by sturdy oarsmen who would row it out in front of the *Lee*, with leadsmen aboard to measure the river's depth with poles and weighted lines. Cannon then stationed himself forward on the hurricane deck, where he could watch and hear the warnings of the men on duty below and immediately relay those warnings to Jameson and King in the pilothouse.

"In the engine room," Wiest, the eyewitness, said, "they were using a moderate head of steam, with a man at the throttle of each engine. If the cry came, 'Hard a starboard!' [signaling a sharp turn to starboard], the pilot stopped a wheel until the boat headed away from the [left] shore, and used the same tactics reversed if the shore was too close on his left."[4]

Cannon was struggling with himself as well as the fog, trying to decide whether his obsession with getting to St. Louis first was driving him, his boat and all aboard to destruction. He was trying to decide whether it was time to call a halt, even if it meant that Leathers and the *Natchez* might pass him. For all he knew, the *Natchez* might not be in fog. Vacillating between stopping and continuing, he at one point told pilot George Clayton he wanted to put into shore, but Clayton apparently talked him out of it, reminding him that fog is not all encompassing, that it diminishes and has an end to it somewhere. And so the creeping *Robert E. Lee* continued blindly upriver, guided by the calls of the leadsmen in the yawl. Even so, it struck the riverbank once, then backed off and found the channel again.

Somewhere around Muddy Island, at about two o'clock in the morning, a breeze sprang up, and gradually the blinding mass of gray tore apart and melted into thin, blowing wisps, drifting aimlessly over the water. By the light of the moon, the *Lee*'s crew could actually see the water. The fog was gone. Up ahead, the big river was wide and clear. In the *Lee*'s pilothouse the pilot rang for the engine room to resume full speed as the yawl turned back to the steamer and the oarsmen and leadsmen climbed aboard. Captain Cannon and his pilots could then see the lights of the community of Grand Tower, which stood on the Illinois shore opposite the rocky islet called the Grand Tower. The *Robert E. Lee*, steaming swiftly, was now just a little more than a hundred miles from St. Louis.

At Grand Tower a reporter for the *St. Louis Republican* issued terse reports on the race from his perspective:

GRAND TOWER, July 4, 1:50 A.M.—The R.E. Lee is now passing at speed unprecedented within the memory of the oldest inhabitant.

2:08 A.M.—The Lee is out of sight, abreast of Devil's Oven. Nothing has been seen of the Natchez yet.

GRAND TOWER, 4:30 A.M.—Fog so dense that no boat can get through it. Nothing heard of Natchez yet.

The fog behind the *Lee* had unpredictably blown away and then had returned, thicker than ever, all to the disadvantage of the *Natchez*. As it remained tied up at Shepherd's Landing, some of the exhausted crew members took as much rest as they could get. Others took turns standing watch, eager for the first sign that the fog was diminishing. After four hours of idleness, just before five o'clock on the morning of July 4, Captain Leathers, thinking he detected a thinning of the fog, ordered the engine room to crank up the *Natchez*'s engines and get it going again. After moving only a few yards, Leathers could see he was mistaken. The fog had yet to break. He ordered the boat tied up again.

Another hour and a half passed. Now it became clear that the fog was indeed dissipating. As the fog broke up, the *Natchez* started up once more, churning its way out into the middle of channel and steaming northward again. Whatever hope Leathers still had for catching the *Robert E. Lee* must have vanished like the fog when the *Natchez* reached Grand Tower and someone in a small boat rowed out to tell him that the *Lee* had passed Grand Tower at two A.M. and was six hours ahead of him.

About five o'clock the sun had come up, wiping out the last remnants of the fog, pouring light onto the wooded riverbanks and brightening the summer sky. It was going to be a glorious Independence Day on the river. On board the surging *Robert E. Lee* Captain Cannon, peering far astern, could see no sign of the *Natchez*, no telltale smudges of black smoke rising above the horizon. It was obvious to him that he and the *Lee* were far, far ahead.

Past Ste. Genevieve the *Lee* steamed, past old Fort de Chartres on the Illinois shore. By mid-morning it had passed Herculaneum, moving now at less than full speed, Cannon and his pilots knowing there was no longer need for it. At the Missouri village of Sulphur Springs, passengers aboard the *Lee* could see an excursion train of the Iron Mountain Railroad stopped on the tracks that ran beside the river, its passengers having left the train to stand on the riverbank to await the *Lee*'s arrival and greet it with waving handkerchiefs and cheers as it steamed past them, swiftly headed for the finish line.

Certain victory was now mere minutes away.

• 14 •

Celebration in St. Louis

When the *Robert E. Lee* approached the lower end of Jefferson Barracks, more cheers of greeting were shouted across the water by passengers on the steam ferryboat *East St. Louis*, which was tied to the riverbank to provide a reviewing stand for a multitude of spectators. As the *Lee* steamed past Jefferson Barracks it received a three-gun artillery salute from the old Army post, and the *Lee* returned the salute with its signal cannon.

At the northern end of Jefferson Barracks four excursion steamers, filled with celebrants, stood in the river waiting for the *Lee*, and when it appeared, the crowds on the steamers and those on the bluffs above Jefferson Barracks roared their greetings and congratulations. Adding to the tremendous welcome was the excursion train that was now dashing toward St. Louis abreast of the *Lee*, blowing its whistle in ear-splitting salute, as the *Lee*'s pilots repeatedly answered with the steamboat's whistle, filling the sultry air with sounds of raucous celebration. The cheering mass of spectators along the shore grew thicker as the *Lee* reached Carondelet and proceeded on toward downtown St. Louis. The scene was recorded in an eyewitness account by a reporter for the *St. Louis Democrat*:

> Long before the Lee was expected, the people began to assemble on the wharf boat, the steamers and the houses fronting the levee. Every boat was crowded with anxious spectators, men, women and children, all determined to see the boat when she came in....
>
> At ten minutes past 11 o'clock a cry was heard from the crowd standing on the levee at the foot of Market street—"There she comes!"
>
> The cry was caught up by the people on the steamers and wharfboats, and along the whole line of the levee it was passed, and "There she comes!" was heard from Market street to the shot tower. There was a movement of the crowd, and all eyes were turned down the river, but no boat was in sight. Then voices exclaimed, "That's a sell!" and the people settled down again, watching with anxious eyes for the first faint cloud of smoke that might appear in the distance. A few minutes passed, and the word was given that she was "Coming sure!"

Four. The Outcome

View of the St. Louis riverfront as it appeared in 1871, a year after the race. The *Robert E. Lee*'s arrival on July 4, 1870, was greeted by a cheering mass of spectators along the shore, by passengers on excursion steamers waiting in the river and by the ear-splitting whistle of a train dashing beside the river, abreast of the *Lee* (Library of Congress).

A tug now put out, and proceeded to the foot of Chouteau Avenue, meeting the Lee near that point. At 11:20 the black smoke from the Lee's chimneys was seen rolling past the foot of Cedar street. A minute later and she shot into view at Market street, and the boom of cannon announced her arrival. The multitude held their breath in eager expectancy, while ever and again the voice of the cannon proclaimed the progress of the victorious steamer. An old colored fireman shouted: "I golly! dat puts me in mind of war times!"

At twenty-five minutes past 11 o'clock the Robert E. Lee passed her place of mooring, at the New Orleans wharfboat, firing a gun as she got opposite. Now the pent-up enthusiasm of the people broke forth in shouts and yells, waving of handkerchiefs and tossing of hats. The salute was returned by those on board the Lee. As she passed we observed two colored men sitting aside the cross-timbers of her jackstaff; one seemed to be playing the banjo, while the other was yelling at the top of his voice....

The Lee passed on to the head of Bloody Island, where she rounded to. The multitude followed her up the levee, and there was a scene of the wildest confusion — men, women and children hurrying along as in chase of the boat; baggage wagons and hotel coaches dashing through the crowd — people rushing on shore from the steamers and wharfboats, and everybody panting with excitement.

The shouts were not so loud as it was expected they would be from the size of the crowd. The enthusiasm was not so demonstrative as became the importance of the occasion. The colored boatmen, however, made a great deal of noise, some of them yelling at the top of their voices, some falling on the ground and rolling over in the agony of delight.

Slowly the victor backed about, and gracefully moved down stream, then rounded to again and came alongside of the wharfboat, to which she was made fast. Then there was a grand rush on board, and the friends of the officers grasped their hands, and tendered their congratulations.[1]

14 • Celebration in St. Louis

The victorious *Robert E. Lee* had reached its destination at 11:24:30 a.m., and its official time from New Orleans to St. Louis was announced to the cheering crowd on a long, canvas banner tied to the larboard railing on the boat's boiler deck:

> N.O. TO ST. LOUIS, 3D., 18H. 14M

The *Lee* had not only beaten the *Natchez* to St. Louis but it had also beaten the record time for a voyage from New Orleans to St. Louis, set by the *Natchez* on June 22, 1870, less than two weeks earlier, by three hours and forty-four minutes.

In New Orleans the feat of the *Robert E. Lee* and its captain and crew was extravagantly heralded on the front page of the *Picayune*, whose readers were informed:

> The greatest steamboat race that has ever taken place on the Father of Waters is over, the hopes and fears of thousands are settled, and the R.E. Lee wears proudly the title of CHAMPION OF THE MISSISSIPPI RIVER.
> Not easily were her honors won, for her rival was swift of keel, and so closely contested the race that it may almost be said she shared the honors with her.
> The people of the entire Mississippi Valley have been excited about this race as they never were before by any similar event, and the banks of the great river were thronged with thousands of deeply interested spectators during the progress of the race, all along the route from New Orleans to St. Louis.
> It is not to be denied that the illustrious name which the victor bears had much to do with the popular sympathy for her in this contest. To such an extent was this feeling carried that we heard of parties who had their money staked on the Natchez declare they would prefer to lose it rather than the Rob't E. Lee should be defeated.[2]

On reaching downtown St. Louis, Captain Cannon had sped the *Lee* past Walnut Street, where it was to land, and as if taking a victory lap, continued up to where the piers for the new bridge across the Mississippi were under construction, then had made a sweeping turn and headed back to Walnut Street, slowed his vessel and tied it up to the wharfboat there. Once it was tied up, the throng of well-wishers pushed their way onto the boat to congratulate all who were aboard, creating a lively celebration.

"On board the *Lee*," the *Democrat*'s reporter wrote, "the scene was one the like of which is seldom witnessed. Although the police placed on the steps leading to the cabin were active and determined, such crowds passed up and through the cabin that hardly anything could be heard for the noise arising from the confused movements."[3]

Cannon found himself swamped by the crowd, but managed to push free of the mass of bodies and make his way off the *Lee* and onto the wharfboat, where he was met by the official welcomers, including Captain Nat Green, who led him away from the throng and into a private office to escape the crowd and confusion, which he seemed to tolerate well enough. "He does not seem exhausted by the vigils necessary for the task performed," the *Democrat* reporter commented. Among the dignitaries on hand to congratulate Cannon were a host of fellow steamboat captains as well as Mary Lee, the thirty-five-year-old daughter of the man for whom the triumphant *Robert E. Lee* had been named, and James B. Eads, designer of St. Louis's new bridge.

A man of forceful personality and opinions, Eads volunteered to Cannon that he would bet a thousand dollars that if the *Lee* had an iron hull, instead of its wooden hull, it would have made the trip from New Orleans faster by five hours. He told Cannon that an iron hull would have made the *Robert E. Lee* a foot lighter in the water and he then pressed Cannon to tell him how much faster the *Lee* could have gone if its draft had been a foot less. Fortunately for Cannon, another well-wisher was brought to him to be introduced then and he was able to turn away from the argumentive Eads.

Captain Cannon did take time to answer other questions, though. One of his fellow steamboat captains asked him about the stage of water he preferred when attempting a fast trip. Cannon quickly responded, "Bank full of water. I want it bank full, always for my fast trip."[4] Cannon, the reporter observed, seemed "very happy," and when asked about his feelings, Cannon replied that if he seemed happy, it was because he had met so many friends and was deeply gratified by the reception given him. Cannon attributed his success to the *Robert E. Lee*'s machinery, calling its engines "the best in the world" and claiming that except for the water leak, the boat's machinery was in as good condition at the end of the race as it was when the *Lee* left New Orleans. Commenting on the fog that had slowed down the *Lee* and had critically delayed the *Natchez*, Cannon admitted, "Someone aboard was in favor of laying up," apparently referring to himself, "but I persisted in running slow, and in a few minutes the fog was left behind."[5]

Amid the hubbub, one of the *Lee*'s passengers, feeling effusive over the success of the *Lee* and Captain Cannon's handling of the vessel, penned a note of gratitude and commendation to him:

> We, the undersigned passengers of the Robt. E. Lee, take this method of tendering our thanks to Capt. John W. Cannon and his officers, for the pleasant trip just made, and would compliment Captain Cannon on his superior judgment and skill in the management of his boat, making the time quicker than it was ever made before. And we must say in praise of the noble craft, that everything worked to the satisfaction of all aboard. And we would hardly have known that she was on a fast trip had it not been for the continued cheering that

greeted us at every landing as we passed. There was no excitement exhibited by the officers and crew during the whole trip. We would say to those who wish to take a pleasant, safe and speedy trip, to go on the "Robt. E. Lee."[6]

The note's author then asked his fellow passengers if they would like to sign the statement, and the thirty who affixed their signatures to the note thereby entered their names into the annals of American maritime history, forever identified as participants in the great river's greatest race.[7]

It was almost six o'clock that evening when the *Natchez* came steaming into sight at St. Louis. As it passed Carondelet, steamers standing in the river greeted the *Natchez* with their whistles and bells, and the crowds on shore, standing on the riverbank and on the porches and balconies of houses, shouted and waved handkerchiefs. The crowds in downtown St. Louis, still celebrating the Fourth of July and the conclusion of the historic race, likewise cheered and hailed the late-arriving *Natchez* as loudly and enthusiastically as they had the *Lee*. As his vessel came up to its wharfboat Captain Leathers again pulled his new watch from his pocket and consulted it. It said 5:51 p.m., New Orleans time. The clock on the wharfboat, however, said 6:02, St. Louis time. In either case, the *Natchez* had finished the course some six and a half hours behind the *Robert E. Lee*.

Warmly greeted by a legion of friends, Leathers also faced the newspaper reporters and others in the crowd who had questions for him. He promptly let them know that he believed the *Natchez* had run a faster race than had the *Lee*. He conceded that the *Lee* had arrived six and a half hours before him, but maintained that allowances should be made for the difficulties the *Natchez* had encountered. He said thirty-six minutes should be subtracted from his boat's running time for the time it lost when it had to stop at Milliken's Bend for repairs to the valve of its intake pump, and more than six hours should be subtracted for the idle time the *Natchez* had spent waiting for the fog to lift. When all was considered, Leathers figured, the *Natchez* had actually made the trip in twelve minutes less than had the *Robert E. Lee*.[8]

Apparently no one in the crowd wanted to argue the matter with the formidable-looking captain of the *Natchez*. "The expression of his countenance," one newspaperman reported, "is open, frank and rather pleasing. But if anyone is willing to calmly read that face, such a one would very probably conclude that he would not like to have him [Leathers] for an enemy."[9]

To the *St. Louis Republican* reporter who had been aboard the *Natchez* all the way from New Orleans and had become one of its champions, Leathers's argument made perfect sense. "The *Natchez*," he wrote, "was beaten to St. Louis several hours, yet if an accurate deduction of the time she lost by accident to her pump and also by making two special landings for passengers alone, together with the time lost in the fog and by her numerous backings toward New Orleans from shoal water [were made], it will appear that her real running time to St. Louis is not greater than that of the *R.E. Lee*."[10]

That same reporter, though, decided that the *Natchez* was no faster in the water than was the *Lee*, that they were equal. "The *Natchez* cannot possibly pass the *Lee* under way. She can get just so close as to ride on her swells and not another inch can she gain. The same would be the case with the *Lee* in the wake of the *Natchez*.... Were they let loose at New Orleans together on a big river they both would reach St. Louis in three days and twelve hours from New Orleans."[11]

Captain Leathers, in an interview with a *St. Louis Democrat* reporter the next afternoon, July 5, remained steadfast in his belief that the race had not proved the *Lee* to be the superior vessel:

REPORTER: "Captain, are you prepared to admit that the Lee is faster than the Natchez?"
LEATHERS: "No. The Lee is not faster, by a long sight. No, sir."
REPORTER: "Any objection to tell me something about your trip?"
LEATHERS: "No. We went to New Orleans and there were over 90 people's names on the register for the Natchez and we were to take passengers at Vicksburg, Greenville and Memphis. We had 40 deck and cabin passengers for Cairo, whom we put on boats or tugs, and we brought through to St. Louis about 70 cabin passengers."
REPORTER: "How about the Lee?"
LEATHERS: "She did not land alongside any wharfboat on the river. We lost thirty-six minutes at Buckhorn in Milliken Bend, and put out twenty passengers at Memphis."
REPORTER: "Were you as thoroughly stripped as the Lee?"
LEATHERS: "We did no stripping except of the extra cattle dunnage, and my boat is in perfect order in every particular."
REPORTER: "What is the fastest trip you can make from New Orleans to St. Louis?"
LEATHERS: I can come in 3 days 12 hours; but I am sure not to try it in shoal water. I think the Natchez has the capacity to do it. But for two of my stoppages, I would have beaten the Lee to St. Louis."
REPORTER: "Wherein was the Lee's greatest advantage in this contest?"
LEATHERS: "She received one hundred cords pine wood off the Pargoud; that was her great aid and advantage, and then I lost six hours in the fog, and the thirty-six minutes I have mentioned before. But for those we would have beaten the Lee's time to St. Louis some twenty odd minutes. My losing landings were at Buckhorn and Devil's Island."
REPORTER: "Then, as to your preparation for a race, captain?"
LEATHERS: "I made none. I took fuel at the usual places, and had assistance from nobody. No fuel but Pittsburg coal."
REPORTER: "Your passenger receipts must be considerable."
LEATHERS: "We had $3000 or $4000 passenger receipts. I wasn't prepared to tear up my boat, but to carry passengers."

14 • Celebration in St. Louis

REPORTER: "How did you expect to get along above Cairo?"
LEATHERS: "I expected to clean her out in this river. At Cape Girardeau I was only one hour behind her. I touched bottom twice, having missed the channel twice. We merely bumped, and immediately backed off."

Here the interview ended, the reporter informing Capt. Leathers that he supposed the Natchez would beat the Lee's time within six months, but he [the reporter] would not ask any information on that point.[12]

Leathers's friends and backers in Cincinnati were generally more gracious, conceding the *Natchez* had been fairly beaten. The Cincinnati *Gazette*, perhaps speaking for them, expressed its feelings in a straightforward, no-excuses editorial published in a late edition on July 4:

> The three days' agony is over. We are glad of it. There can be no doubt as to which is the fleetest steamer on the Mississippi. The Robert E. Lee need not make another run until a steamboat is built in the future that, upon trial, will excel her in speed. Cincinnati may be proud of the Natchez for her beautiful model. Her machinery is also good, else she would not have made the run she did — sometimes even gaining on the Lee.
>
> "Generalship" and many other things may have had their influence on the race, but the solid fact stares us in the face that the Lee has beaten the Natchez. The reason of this, to plain, common-sense people, is apparent, namely: The Lee is the fastest boat.
>
> The Natchez was built expressly to beat the Lee. The question heretofore has been, "Will the Lee beat?" The only question now is, "Has she done it?" We think she has, and fairly, too. The thirty-four-inch cylinders could not cope with the forty-inch cylinders. Cincinnati must build another boat and try the large cylinders.[13]

By the next evening, Tuesday, July 5, the officers of the two steamboats were sufficiently recovered from their ordeal to attend the banquet planned to honor the loser as well as the winner. The celebration of the consummation of the great race was to be held at the Southern Hotel, four blocks back from the riverfront, on Walnut Street, a hotel widely known for excellent cuisine. The banquet's organizing committee had engaged Postlewaite's String Band to provide entertainment for the fifty or so invited guests, all men, most of them steamboat captains and officers. A long-time captain, Dan Taylor, apparently picked for his gift of eloquence, was to preside over the affair.

In the hotel's banquet room the guests took their places at three long, parallel tables that awaited them. At the head of the table on the right was seated the St. Louis harbormaster, Captain R.P. Clark, and beside him sat John Cannon. The other officers of the *Robert E. Lee* sat on either side of the table, along with Cannon's old partners, Johnny Smoker and John Tolle, and others. It was obviously the *Lee* table, with a model of the vessel set on it as a centerpiece. One of those seated at it, however, was N.C. Claiborne, who was related to Leathers by marriage.

At the head of the table on the left sat Captain W.W. Green. Tom Leathers sat to the right of him, and the other officers of the *Natchez*, as at the *Lee* table, were arranged on both sides of the table. Also at the table were, among others, Bart Able, captain of the *Mollie Able*, and John Christy, who had traveled from Memphis to be with his friend Tom Leathers at the celebration. At the center of the table stood a model of the *Natchez*.

Between the *Lee* and *Natchez* tables was the center table, which may have been regarded as the neutral, or barrier, table. At the head of it sat — and occasionally stood — the banquet's master of ceremonies, Captain Taylor, and arranged on either side of it were the rest of the celebrants.

The food and wine were as rich and bountiful as might be expected aboard a grand Mississippi River steamer, and Postlewaite's String Band performed superbly, according to all reports. When the dinner had been consumed, wine glasses emptied many times, cigars lighted and Postlewaite's musicians had been stilled, Captain Taylor rose from his chair at the center table and began an oration of complimentary remarks, followed by a reminder of what the guests had come to celebrate. "The two steamers that arrived at our wharf yesterday," he said, revealing the public-relations aspect of the race, "have demonstrated that steamers can navigate even the difficult waters of the Mississippi and yet compete with the railroad, as they have done."

His audience applauded in hearty agreement.

"There are many of you here," Taylor went on, "who have been longer upon the railroad from New Orleans to St. Louis than any of the gentlemen who have arrived on these steamers." More applause and shouts of agreement. "I speak of that in general terms, previous to congratulating our friends, the officers of the two steamers who have been so successful and who have so perfectly illustrated the fact that steamers can navigate the Mississippi River and yet compete with railroads on the land, even though they cut across the corners." That statement was intended as a joke, as his listeners understood it to be, and they responded with laughter and more applause.

After a few more remarks in praise of steamboats, Taylor lifted his wine glass and announced, "I now offer you a toast, gentlemen. The crews of the steamboats *Lee* and *Natchez*!"

His audience quickly responded with their own raised glasses and drank his toast. Taylor then called upon Captain Cannon for a few words.

Cannon declined, and Colonel Claiborne of Kentucky, Leathers's relative, stood to speak on behalf of Cannon. "I am related to both of these boats," he told the celebrants, "to the *Natchez* by ties of blood, to the *Robert E. Lee* by state pride." After reciting the achievements of the grand old steamer *J.M. White* as well as the *Natchez* and the *Robert E. Lee*, Claiborne reminded his listeners that the steamboat was invented by an American. "I make this statement," he said, "when I know and feel that we have a better people, braver men and prettier women than there are in any nation." Laughter and thunderous applause followed.

"There is a name, gentlemen, in this celebration," he went on, turning serious. "There is a name upon the card that invited me to this ovation, and there is something in that name, and I beg you to go slow." The room grew quiet. "If any words should pass my lips hastily, that should sound harshly upon the ears of the most sensitive, I will pour upon the wound the balm of a thousand oils before I get through."

His preparations made, he then delivered his tribute to what was in that mixed audience a controversial figure — the Confederate general for whom the winning steamer was named, Robert E. Lee. "The whole people of this great country respect the man, though they condemn his course," Claiborne said. He pronounced himself one of those who "rejoiced in the final victory of the Union" and then, seeking to salve the war's wound, offered a little lightness. "Suppose," he said, "we had divided, where would the *Natchez* and the *Lee* have stopped? Not at St. Louis. And we would have been put out of this banquet tonight, and the wine and good spirits and good cheer we have had."

The audience responded as he no doubt had hoped, with laughter and applause. Finally, he raised his glass and called for, "Long life and good health to Tom Leathers and John Cannon!"

That toast downed, Captain Taylor stood and turned to Leathers for a speech. Like Cannon, Leathers declined. Captain Able of the *Molly Able* rose to speak for him. By now the wine evidently was having some effect on the relevance and the reality of the remarks. Able wanted America to have credit for more than the invention of the steamboat. He pointed out for the audience that an American, Samuel F.B. Morse, had invented the telegraph, and another American, Cyrus W. Field, was the man who had laid the trans–Atlantic telegraph cable. The celebrants, now in a mood to cheer any agreeable statement, cheered for Morse and Field and Able's reminder of their accomplishments. He took time to deplore the recent war and declared himself happy that peace had been reached, allowing the occurrence of such a great event as the running of the race between the *Natchez* and the *Robert E. Lee*. "It is an event," he declaimed, "which has stirred the American heart to its very core. There is no part of this great nation that has not responded to this great race of steam."

He then grandiosely predicted that the attention drawn to the mid-continent by the race and the resulting realization of the vitality of Mississippi River commerce would in the immediate future cause the nation to move its capital from Washington to St. Louis. His audience cheered that also.

"I trust," he said, concluding, "this will be the inauguration of a better, more cordial and social era in the life of western boatmen, and though the railroad car goes and the telegraph flashes on every side, there is no obstacle to shut this mighty river.... Your sons' descendants will yet navigate its great waters and perhaps achieve greater triumphs than those who passed before them."

He then returned himself to his chair. The speeches went on, however, speaker after speaker making fanciful remarks and calling for toast after toast. Everyone had a chance to rise and speak. One of those who spoke raised the memory of the late Captain J.M. Convers, former master of the old *J.M. White*, and at that, Postlewaite's musicians started playing "Auld Lang Syne," and the celebrants sang along. When George Clayton, chief pilot of the *Robert E. Lee*, was asked to make a speech, Postlewaite's String Band broke out into the rousing strains of "Dixie." Clayton, however, begged off the requested speechmaking, pleading exhaustion.

The harbormaster, R.P. Clark, volunteered a few tall tales, one of them being how his baldness was a result of Indians shooting his hair off with muskets. More laughter and cheers. One of the most senior members of the audience, Captain Reuben Ford, stood and boasted to the audience that in his long career on the river he had held every job known aboard a steamboat. "You were not chambermaid!" Dan Taylor rejoined, drawing huge laughs and howls from the crowd.

The toasts became all inclusive — to the citizens of St. Louis, the citizens of New Orleans, the chairman of the banquet committee, the proprietors of the Southern Hotel — until at last the celebrants were ready to call it a night and make their way home or back aboard their vessels.[14] Captain Cannon and his officers returned to the *Robert E. Lee*. Captain Leathers found accommodations on his wharfboat.

Early the next morning, Wednesday, July 6, Cannon and his crew made ready to depart St. Louis, and at eight o'clock the *Lee* drew in its lines and backed away from its wharfboat as a crowd watched from shore. It steamed upstream to the northern end of the waterfront, then turned about, fired its signal cannon, and with a full head of steam, glided swiftly past downtown St. Louis, headed for Mound City, where it would enter drydock and undergo repairs to overhaul its engines and boilers and restore its stripped upper works and have its hull repainted as well.

If Cannon thought Tom Leathers might be planning to stage another race with the *Robert E. Lee*, running downriver this time, he need not have worried. Leathers and the *Natchez* spent July 6 and 7 taking on passengers and freight for the return trip to New Orleans and did not leave St. Louis until the evening of July 7, well after the *Lee*'s departure.

The racing of the *Natchez* and the *Robert E. Lee* had indeed ended.

Epilogue

While the *Natchez* was resuming its service between New Orleans and St. Louis, the *Robert E. Lee* remained in Mound City undergoing repairs and restoration until September 1, then steamed away to New Orleans and on September 20 departed New Orleans, again hailed by a huge crowd on the riverfront, to resume its regular run to Vicksburg, making the first trip of its regular service since the race.

Tom Leathers, still bent on proving the speed of the *Natchez*, on October 16 raced against the *Lee*'s record time from New Orleans to Natchez and beat it by nineteen and a half minutes, winning back the horns. Less than two weeks later the *Robert E. Lee* reclaimed the horns by bettering the *Natchez*'s latest best time by fifteen minutes, making the trip from New Orleans to Natchez in sixteen hours, thirty-six minutes and forty-seven seconds.

On December 21, 1870, the steamer *Potomac* accidentally rammed the *Lee* at New Orleans, staving in its hull and sinking the *Lee*, but without any loss of life. While it was being raised, a fire broke out on the New Orleans riverfront on January 1, 1871, destroying four steamers docked there, but leaving the water-logged *Robert E. Lee* untouched by the new disaster. After it was lifted from the river and refitted, the *Lee* returned to service, still competing with the *Natchez* for business if not in races.

During the cotton season of 1874 the *Lee* on one voyage to New Orleans hauled a load of 5,741 bales aboard its decks, surpassing the record load of five thousand bales carried by the *Natchez* on a trip in 1872.

By 1874, Captain Cannon's eldest son, William, twenty years old in August of that year, had joined the crew of the *Lee*.

In 1876 Cannon took the *Robert E. Lee* up the Ohio River to Jeffersonville, Indiana, opposite Louisville, on its final voyage. The superstructure was stripped and its parts disposed of, some of the elegant chandeliers from its saloon being donated by Cannon to the Presbyterian church in Port Gibson, Mississippi, and its trophies being transferred to the *Lee*'s successor, the *Robert E. Lee II*, larger and even more luxurious than the vessel it replaced. The worn-out hull of the old *Lee* was towed to Memphis, where it served out the remainder of its useful years as a wharfboat. The new *Lee* was launched on April 25, 1876, with William Cannon as its clerk and John Cannon as its captain.

After ten years of service, the *Natchez* was also ready for replacement. In June 1879 Captain Leathers took it on a voyage to Cincinnati, where its successor, the seventh *Natchez*, was being built. On the way, it ran aground on a sandbar and despite all its efforts and the help of tugboats, it could not be dislodged. "It would be a damn sight more romantic for the old craft to die and be dismantled midstream," Leathers remarked almost wistfully, "with years rippling around her, and not in the boneyard." Then, perhaps thinking of the hazard to navigation the abandoned steamer would present—and of his liability for it—he had a quick second thought. "But," he said, "it's too damn troublesome."

And so Leathers and his crew left the grand old steamboat stuck on the sandbar and waited till the river at long last rose and lifted the *Natchez* free. Once refloated, it was stripped and dismantled and the remaining hulk was sold for two thousand dollars. Like its former competitor, it became a wharfboat, permanently moored at the Refuge Oil Mill, on the Mississippi River below Vicksburg.

The glorious old racers had finished their last course.

A group of steamboat owners in St. Louis organized a corporation to consolidate their assets and strengths and chose Cannon to be its chief executive. He never lived to take the job, though. Plagued by a series of colds and poor health but unwilling to alter his schedule or work habits, Cannon contracted pneumonia and died at his home in Frankfort, Kentucky, on April 18, 1882 at age sixty-one. His body was buried in Frankfort. E.W. Gould, one of Cannon's fellow steamboat captains, summed up the life of the gallant old steamboatman:

> Laudable ambition was his peculiarity. Honesty and integrity marked his course through life. Kindness, generosity and suavity were prominent virtues in his character.
> His great ambition to excel all competitors involved his health and his fortune. And although a man of remarkable physique and good judgment, his ambition probably destroyed both.[15]

Tom Leathers continued to operate the seventh *Natchez* and later the sternwheeler *T.P. Leathers,* but ran into bad luck with both. The hull of the seventh *Natchez* sprang a leak at Stack Island in the Mississippi and sank on

New Year's Day 1889. In November 1890 the *T.P. Leathers*, loaded with 1,700 bales of cotton and 8,757 sacks of cottonseed, also sank, about three miles above Natchez. He then built another *T.P. Leathers* and another *Natchez*, but turned the running of them over to his sons Frank and Bowling. The old captain remained a partner in the firm of Leathers and Hoey, steamboat agents in New Orleans, and took another son, Tom Jr., into the firm with him.

On the evening of June 1, 1896, a week after celebrating his eightieth birthday, Leathers set out for a walk from his big brick house at the corner of Carondelet and Josephine streets in New Orleans and as he was crossing St. Charles Avenue, one block from his house, he was struck and knocked to the ground by a bicycle speeding through the darkness. Bystanders carried him back to his house, where he died twelve days later, on June 12, 1896. His body was buried in the city cemetery in Natchez.

A eulogy by the New Orleans *Daily States* marked his passing. "There are many who regarded Capt. Leathers as the greatest of steamboaters," it said with carefully chosen words. "Certainly no captain, in the history of the river, achieved greater success, was more widely known or more highly respected, and few men ever presented such a picturesque and commanding appearance."[16] At the funeral, the officiating minister, the Rev. Dr. B.M. Palmer, saw historic significance in the end of Leathers's life. "He is one whose death," the minister declared, "is like the death of the century."[17]

Death for the Mississippi River steamboat itself was not long to follow. Ever since the Charleston & Hamburg line of South Carolina had run the first steam locomotive in December 1830, followed by the Baltimore & Ohio in the summer of 1831, railroads had been spreading across the country like vines. In 1835, just five years after the Charleston & Hamburg had carried some two hundred passengers on its historic first steam-locomotive run, there were 1,098 miles of track upon which steam railroads were operating in the United States. By 1840 there were an estimated 3,000 miles of track. Only four of the nation's twenty-six states had no tracks laid by 1840—Vermont, Tennessee, Missouri and Arkansas. By early 1837 at least two hundred railroads were either already in operation or were being built, planned or being considered.

The United States postal service quickly saw the possibilities for moving mail by railroad. By 1834 it was using trains to send batches of mail in pouches. In 1838 the U.S. Congress enacted a law making all railroads postal routes, and having the mail sped along by rail became an ordinary occurrence.

By the outbreak of the Civil War in 1861 most of the country, particularly east of the Mississippi, was laced with railroad lines. The numbers revealed the trend. In 1850 total track mileage in the United States was 9,000 miles (up from 3,000 miles in 1840). By 1860 the total had risen to 30,000 miles. In 1870 the total was 53,000 miles, and by 1880 it had swelled to 93,000 miles and was still growing. The crowning achievement of the railroad builders came on Monday, May 10, 1869, when, in an act that was both the symbol and the deed of the railroad's conquest of America, the Central Pacific and the Union Pacific railroads met at Promontory, Utah, and joined their tracks into a transcontinental rail route that stretched from New York to California.

Over time, a great deal of effort was put into making railroad passengers more comfortable. The cars' interiors were decorated to resemble hotel rooms, with curtains, upholstered seats and varnished or painted woodwork. The most elaborate cars came to resemble ornate Victorian parlors, and by the 1860s passenger cars came equipped with toilets. In 1863 George M. Pullman, a cabinet and coffin maker turned building contractor and inventor, patented a sleeping car with upper berths that folded out to make a bed and, below them, seats that could be extended to form a lower-berth bed, all of his invention. In 1867 Pullman introduced another revolutionary innovation, a sleeper car to which was attached a car that was a rolling restaurant, with a compact kitchen and a gracious dining room included.

Trains then became hotels on wheels, and railroads sped into a whole new era of transportation, one in which steamboats became a dangerously threatened species.

For a time, showboats helped keep the Mississippi River steamboat a presence in the lives of people in communities along the river, even while railroads were thinning out the number of packets on the Mississippi. The steamboat had been adapted as a floating theater as early as 1836, when the Chapmans — a nine-member family of traveling actors — bought their first steamer and took it and their performances to communities on the Mississippi and its tributaries. Later showboats, some of them towing barges on which arenas had been built, were little more than floating circuses, with extensive menageries of exotic animals. The showboats lasted into the twentieth century. One, the *Goldenrod*, was operated by a succession of owners through the 1980s. In the 1990s it was renovated and operated as a dinner theater, docked at St. Charles, Missouri. It still survives as a National Historic Landmark, a museum piece, the last of the old-time Mississippi River showboats.

By 1875 it was obvious the Mississippi River steamboat was in its death throes. "The direct and immediate cause for the great decline in this important branch of commerce," the former captain and steamboat historian E.W. Gould publicly complained in January 1875, "is, of course, the construction of so large a number of railroads." What was *not* being constructed then, he pointed out, were steamboats. Whereas in the years shortly prior to 1874, an average of one hundred new steamers were built each year, Gould wrote in the *Nautical Gazette*, "in 1874 there was but a single boat built of any considerable capacity, of the usual kind, for freight and passengers, and but very few tow-boats, or any other character of [steam] boat."[18]

With the spread of railroads, which could transport passengers and freight faster, cheaper and to more destinations than could river-bound steamboats, the public's demand for steamboat service had simply vanished. Samuel Clemens charmingly captured the turn of events in his *Life on the Mississippi*, written in 1883, quoting from his conversation with an old-time steamboat clerk:

> "Boat used to land — captain on hurricane roof — mighty stiff and straight — iron ramrod for a spine — kid gloves, plug tile, hair parted behind — man on shore takes off hat and says:
>
> "'Got twenty-eight tons of wheat, cap'n — be great favor if you can take them.'
>
> "Captain says: 'I'll take two of them' — and don't even condescend to look at him.
>
> "But nowadays the captain takes off his old slouch, and smiles all the way around to the back of his ears, and gets off a bow, which he hasn't got any ramrod to interfere with, and says:
>
> "'Glad to see you, Smith, glad to see you — you're looking well — haven't seen you looking so well for years — what you got for us?'
>
> "'Nuth'n', says Smith, and keeps his hat on and just turns his back and goes to talking with somebody else.
>
> "Oh, yes! Eight years ago the captain was on top; but it's Smith's turn now. Eight years ago a boat used to go up the river with every stateroom full, and people piled five and six deep on the cabin floor; and a solid deck-load of immigrants and harvesters down below, into the bargain.... But it's all changed now; plenty staterooms above, no harvesters below.... they've gone where the woodbine twineth — and they didn't go by steamboat, either; went by the train."[19]

Clemens lived long enough to see the end of the steamboat era — and to lament it. "Mississippi steamboating was born about 1812," he wrote, as if penning its obituary; "at the end of thirty years it had grown to mighty proportions; and in less than thirty more it was dead! A strangely short life for so majestic a creature."[20]

Truly it was.

Chapter Notes

Introduction
1. E.W. Gould, *Gould's History of River Navigation*, pages 527, 529.

Chapter 1
1. Mark Twain, *Life on the Mississippi* (New York: Harper & Row), pages 100–101.
2. Herbert Quick and Edward Quick, *Mississippi Steamboatin'*, page 207.
3. New Orleans *Daily Picayune*, July 2, 1870, page 1.
4. *The New York Times*, July 2, 1870.
5. Julian Street, *American Adventures*, page 513.
6. According to Henry Clay Warmoth, in his book, *War, Politics, and Reconstruction*, pages 157–158, the *Robert E. Lee* was financed and owned by two Northern businessmen, Oakes Ames of Massachusetts and Ames's representative in New Orleans, Asa S. Mansfield. I found no other source to confirm Warmoth's claim, which includes his statement, "But only a few knew that the 'Lee' was owned by Mansfield and Ames."
7. Frederick Way Jr., "Cannon Took a Poke at Leathers," *Waterways Journal*, July 31, 1943.
8. Henry Clay Warmoth, *War, Politics, and Reconstruction*, pages 158–159.
9. These specifications are attributed to the *Directory of Western River Packets*, by Frederick Way Jr. and quoted in Manly Wade Wellman's *Fastest on the River*, page 186.
10. This quote is reconstructed from the account given to Alfred Pirtle by John Wiest and published in the Louisville *Courier-Journal* in 1916, an undated clipping of which is in the New York Public Library Science and Technology section.
11. This time is reported in *The Great Steamboat Race Between the Natchez and the Robert E. Lee*, by Roy L. Barkhau, page 21. Other sources vary.
12. This time is from Barkhau. Other sources vary.

Chapter 2
1. Francis Parkman, *LaSalle and the Discovery of the Great West* (Boston: Little, Brown, 1903), page 308.
2. Ibid., page 35.
3. Thomas Fleming, *The Louisiana Purchase*, pages 37–38.
4. John Kukla, *A Wilderness So Immense*, page 6.
5. Mark Twain, *Life on the Mississippi*, page 233.
6. Ibid., page 228.
7. George P. Kelley, "Mouth of Arkansas — Napoleon," www.rootsweb.com/~ardesha/napoleon.htm.

Chapter 3
1. "The Great Race," *Picayune*, July 1, 1870, page 1.
2. *St. Louis Republican*, July 6, 1870, page 1.
3. Details of the repair surmised in Manly Wade Wellman, *Fastest on the River*, pages 21–22.
4. Warmoth, page 160.
5. Details of the boiler leak's repair, possibly exaggerated but considered plausible, were provided by John Wiest in a newspaper interview published in 1916 in the Louisville *Courier-Journal* and archived as an undated

clipping in the New York Public Library Science and Technology section.
  6. *Picayune*, July 6, 1870, page 2.
  7. *St. Louis Republican*, July 6, 1870.
  8. Ibid.
  9. Memphis *Appeal*, June 26, 1870.
  10. *Picayune*, July 3, 1870.
  11. *St. Louis Republican*, July 2, 1870.
  12. *Picayune*, July 6, 1870.
  13. Ibid.
  14. Ibid., July 2, 1870.
  15. Street, page 154.
  16. George Devol, *Forty Years as a Gambler on the Mississippi*, pages 239–240.
  17. *Picayune*, July 3, 1870.
  18. Ibid.

Chapter 4

  1. S.C. Gilfillan, *Inventing the Ship*, page 73.
  2. Ibid., page 74.
  3. Fulton letter to Livingston dated June 13, 1802, Clermont State Historic Park. Quoted in Cynthia Owen Philip, *Robert Fulton, A Biography*, page 131.
  4. Kirkpatrick Sale, *The Fire of His Genius*, pages 87–88.
  5. James Flexner, *Steamboats Come True: American Inventors in Action*, pages 291–292.
  6. Ibid., page 119.
  7. *American Citizen*, August 17, 1807. Quoted in Philip, page 199.
  8. From the account that appears on pages 202–203 of *Robert Fulton and the Clermont* (New York: Century, 1909), by Alice Crary Sutcliffe, Fulton's great-granddaughter.
  9. Philip, page 202.

Chapter 5

  1. Lydia years later wrote a letter to E.W. Gould recounting her experiences on the voyage. Part of the letter is reprinted in Gould's book, *Gould's History of River Navigation*, pages 87–89.
  2. Gould, page 88.
  3. Ibid., page 89.
  4. Ibid., page 97.
  5. Ibid.
  6. Ibid., pages 88–89.
  7. Ibid., page 84.
  8. Henry Howe, *The Great West*, page 241.
  9. Gould, pages 98–99.

Chapter 6

  1. Herbert Quick and Edward Quick, *Mississippi Steamboatin'*, page 89. Other sources vary.
  2. Quoted in Florence L. Dorsey, *Master of the Mississippi*, page 111.
  3. Quoted in Adam I. Kane, *The Western River Steamboat*, page 50.
  4. Quick and Quick, page 92.
  5. Dorsey, page 128.
  6. Samuel Treat, "Political Portraits With Pen and Pencil: Henry Miller Shreve," *The United States Magazine and Democratic Review* volume 22, 1848.
  7. Quoted in Dorsey, page 138.

Chapter 7

  1. Fred Erving Dayton, *Steamboat Days*, page 92.
  2. Ibid., pages 93–94.
  3. Ibid., page 108.
  4. Quick and Quick, pages 170–171.
  5. Erick F. Haites, James Mak, and Gary M. Walton, *Western River Transportation, 1810–1860*, page 158.
  6. Quick and Quick, pages 175–177.
  7. Ibid., pages 175–176.
  8. Dayton, page 349.

Chapter 8

  1. Twain, page 218.
  2. Ibid., page 222.
  3. B.A. Botkin, *A Treasury of Mississippi Folklore*, page 334. From "Steamboats at Louisville and on the Ohio and Mississippi Rivers," by Arthur E. Hopkins, *The Filson Club History Quarterly* 17, no. 3 (July 1943), pages 146–148.
  4. George Byron Merrick, *Old Times on the Upper Mississippi: The Recollections of a Steamboat Pilot from 1854 to 1863* (St. Paul: Minnesota Historical Society Press, 1987), page 152.
  5. Quoted from an account written by Frederick Law Olmsted around 1856. Published in Fred Erving Davis, *Steamboat Days*, page 347.
  6. Quick and Quick, page 254.
  7. Ibid., page 128.
  8. Merrick, pages 156–157.
  9. John Morris, *Wanderings of a Vagabond* (New York: self-published, 1873), pages 422–425.
  10. Ibid.

11. Ibid.
12. Ibid.
13. Ibid., pages 140–141.
14. Devol, pages 177–178.

Chapter 9

1. All these statistics are from Thomas C. Buchanan's *Black Life on the Mississippi*, page 10.
2. Quick and Quick, pages 235–236.
3. Buchanan, page 57.
4. Ibid.
5. Frederick Law Olmsted, *Cotton Kingdom*, page 274.
6. Twain, page 72.
7. Buchanan, page 71.
8. Merrick, page 128.
9. Ibid., page 135.
10. Ibid., page 134.
11. Dayton, page 343.

Chapter 10

1. Gould, page 631.
2. Quick and Quick, pages 166–167.
3. Merrick, page 72.
4. Ibid., page 74.
5. Gould, pages 682–683.
6. Ray Samuel, Leonard V. Huber, and Warren C. Ogden, *Tales of the Mississippi*, pages 189–190.
7. Ibid., page 190.
8. Ibid., page 191.
9. Merrick, pages 68–69.
10. Quick and Quick, page 194.
11. Ibid., page 190.
12. Ibid., pages 185–186.
13. Merrick, page 90.
14. Twain, page 160.
15. Merrick, page 57.

Chapter 11

1. Samuel, Huber and Ogden, page 145.
2. Ralph K. Andrist, *Steamboats on the Mississippi*, page 119.
3. Another source says the ship was named the *Trenton*.
4. Gould, page 459.
5. Botkin, pages 294–295.
6. Ibid., pages 124–125.
7. Samuel, Huber and Ogden, pages 123–124.
8. Jerry O. Potter, *The Sultana Tragedy*, page 68.

9. The number of victims of the *Titanic*, which sank in the north Atlantic on April 15, 1912, is also in dispute, the estimates ranging from 1,490 to 1,523.

Chapter 12

1. *Picayune*, July 6, 1870, page 2.
2. *St. Louis Republican*, July 3, 1870, page 1.
3. *Picayune*, July 6, 1870, page 2.
4. Ibid., July 3, 1870, page 1.
5. Ibid., July 6, 1870.
6. Ibid.
7. Ibid.
8. Ibid.
9. Ibid., July 3, 1870, page 1.
10. *St. Louis Republican*, July 4, 1870, page 1.
11. Ibid.
12. Ibid.
13. Ibid.
14. *Picayune*, July 4, 1870, page 1.
15. Alfred Pirtle article in the *Louisville Courier-Journal*, undated clip in New York Public Library Science and Technology section.
16. *Picayune*, July 5, 1870.
17. Ibid.

Chapter 13

1. *St. Louis Republican*, July 6, 1870.
2. John Wiest in interview with Alfred Pirtle, *Louisville Courier-Journal*, undated clip in New York Public Library Science and Technology section.
3. Ibid.
4. Ibid.

Chapter 14

1. From the *St. Louis Democrat*, reprinted in the New Orleans *Daily Picayune*, July 9, 1870, page 1.
2. *Picayune*, July 6, 1870, page 1.
3. Ibid., July 9, 1870, page 1.
4. Ibid.
5. Ibid.
6. *St. Louis Republican*, July 6, 1870.
7. The names of the note's signatories, in the order in which they signed: S.H. Parsiot, A.C. McKeen, Mrs. A.C. McKeen, Miss Maggie McKeen, Francis Shuber, Mrs. F. Shuber, Miss A. Shuber, Mrs. Barry, F. Lonsdale, E.P. Johnson, Edward S. Levy, L.M.

Levy, R.W. Doyle, John Kours, W.L. Calhoun, C. Holmes, Lytle Rowan, J.W. Dougherty, A.L. Long, Albert G. Eberman, J. Kain, Thos. Skinner, J.R. Scanlan, E.M. Jones, John Crozier, J.N. Ryley, G. Williams, R. Frazier, M. Martin, J. Shurad, Alex Warwick of New York.

8. Manly Wade Wellman, *Fastest on the River*, page 131.
9. *St. Louis Republican*, July 6, 1870.
10. Ibid.
11. Ibid.
12. The interview was reprinted in the *Picayune*, July 9, 1870.
13. Editorial quoted in Wellman, page 136.
14. Details of the banquet and quotes of the speakers are from the *St. Louis Republican*, July 6, 1870.

Epilogue

1. Gould, page 725.
2. Quoted in Wellman, page 174.
3. Ibid.
4. Gould, pages 586–587.
5. Twain, pages 322–323.
6. Ibid., page 135.

Bibliography

Books

Ambrose, Stephen E. *Nothing Like It in the World: The Men Who Built the Transcontinental Railroad, 1863–1869*. New York: Simon & Schuster, 2000.
Andrist, Ralph K. *Steamboats on the Mississippi*. New York: American Heritage, 1962.
Barkhau, Roy L. *The Great Steamboat Race Between the* Natchez *and the* Rob't. E. Lee. Cincinnati: Steamship Historical Society of America, Cincinnati Chapter, 1962.
Berry, Chester D., ed. *Loss of the Sultana and Reminiscences of Survivors*. Knoxville: University of Tennessee Press, 2005.
Botkin, B.A., ed. *A Treasury of Mississippi River Folklore*. New York: Bonanza Books, 1978.
Brown, William Wells. *The Narrative of William W. Brown, a Fugitive Slave*. Boston: IndyPublish.com, 2006.
Bryant, Billy. *Children of Ol' Man River*. Chicago: Lakeside Press, 1988.
Bryant, William O. *Cahaba Prison and the* Sultana *Disaster*. Tuscaloosa: University of Alabama Press, 1990.
Buchanan, Thomas C. *Black Life on the Mississippi*. Chapel Hill: University of North Carolina Press, 2004.
Burman, Ben Lucien. *Look Down that Winding River*. New York: Taplinger, 1973.
Cameron, Barbara, and Jerry Stebbins. *Mississippi River: A Photographic Journey*. New York: St. Martin's Press, 1987.
Cooley, Thomas M., et al. *The American Railway, Its Construction, Development, Management, and Appliances*. New York: Arno Press, 1976.
Dangerfield, George. *Chancellor Robert R. Livingston of New York, 1746–1813*. New York: Harcourt, Brace, 1960.
Dayton, Fred Erving. *Steamboat Days*. New York: Tudor, 1939.
Devol, George H. *Forty Years a Gambler on the Mississippi*. Bedford, MA: Applewood Books, 1996.
Dorsey, Florence L. *Master of the Mississippi*. Gretna, LA: Pelican, 1998.
Elliott, James W. *Transport to Disaster*. New York: Holt, Rinehart and Winston, 1962.
Feldman, Jay. *When the Mississippi Ran Backwards*. New York: The Free Press, 2005.
Fichter, George S. *First Steamboat Down the Mississippi*. Gretna, LA: Pelican, 1989.
Fleming, Thomas. *The Louisiana Purchase*. Hoboken, NJ: Wiley, 2003.
Flexner, James Thomas. *Steamboats Come True: American Inventors in Action*. New York: Fordham University Press, 1992.

Gilfillan, S.C. *Inventing the Ship*. Chicago: Follett, 1935.
Gordon, Sarah H. *Passage to Union: How the Railroads Transformed American Life, 1829–1929*. Chicago: Elephant Paperbacks, 1997.
Gould, E.W. *Gould's History of River Navigation*. St. Louis: Nixon-Jones Printing, 1889.
Graham, Philip. *Showboats: The History of an American Institution*. Austin: University of Texas Press, 1951.
Haites, Erik F., James Mak, and Gary M. Walton. *Western River Transportation: The Era of Early Internal Development, 1810–1860*. Baltimore: Johns Hopkins University Press, 1975.
Holbrook, Stewart H. *The Story of American Railroads*. New York: Crown, 1947.
Jensen, Oliver. *The American Heritage History of Railroads in America*. New York: Bonanza Books, 1975.
Kane, Adam L. *The Western River Steamboat*. College Station, TX: Texas A&M University Press, 2004.
Kane, Harnett T. *Natchez on the Mississippi*. New York: Bonanza Books, 1947.
Kukla, Jon. *A Wilderness So Immense: The Louisiana Purchase and the Destiny of America*. New York: Anchor Books, 2004.
Latrobe, John H.B. *Southern Travels: Journal of John H.B. Latrobe 1834*. New Orleans: Historic New Orleans Collection, 1986.
Lucas, Theo., Frank D. Graham, and N. Hawkins. *Audel's New Marine Engineers Guide*. New York: Audel, 1918.
Marquette, Jacques. *Voyages of Marquette*. Ann Arbor: University Microfilms, 1966.
Mason, Philip P., ed. *Schoolcraft's Expedition to Lake Itasca: The Discovery of the Source of the Mississippi*. East Lansing: Michigan State University Press, 1993.
McCague, James. *Mississippi Steamboat Days*. Champaign, IL: Garrard, 1967.
McCall, Edith. *Mississippi Steamboatman: The Story of Henry Miller Shreve*. New York: Walker, 1986.
Merrick, George Byron. *Old Times on the Upper Mississippi: The Recollections of a Steamboat Pilot from 1854 to 1863*. St. Paul: Minnesota Historical Society Press, 1987.
Middleton, Pat. *Discover! America's Great River Road*. Vols. 1 through 4. Stoddard, WI: Heritage Press, 1998, 1999, 2000; Great River Publishing, 2005.
Monjo, F.N. *Willie Jasper's Golden Eagle*. New York: Doubleday, 1976.
Morgan, John S. *Robert Fulton*. New York: Mason/Charter, 1977.
Morrison, John H. *History of American Steam Navigation*. New York: Stephen Daye Press, 1958.
North, Sterling. *The First Steamboat on the Mississippi*. Boston: Houghton Mifflin, 1962.
Philip, Cynthia Owen. *Robert Fulton: A Biography*. New York: Franklin Watts, 1985.
Potter, Jerry O. *The Sultana Tragedy: America's Greatest Maritime Disaster*. Gretna, LA: Pelican, 1992.
Quick, Herbert, and Edward Quick. *Mississippi Steamboatin'*. New York: Henry Holt, 1926.
Ruth, Maria Mudd. *The Mississippi River*. New York: Benchmark Books, 2001.
Sale, Kirkpatrick. *The Fire of His Genius: Robert Fulton and the American Dream*. New York: The Free Press, 2001.
Samuel, Ray, Leonard V. Huber, and Warren C. Ogden. *Tales of the Mississippi*. Gretna, LA: Pelican, 1992.

Stover, John F. *American Railroads*. 2nd ed. Chicago: University of Chicago Press, 1997.
Street, Julian. *American Adventures*. New York: Century, 1917.
Twain, Mark. *Life on the Mississippi*. New York: Harper & Row Perennial Classics, 1965.
Warmoth, Henry Clay. *War, Politics, and Reconstruction*. Columbia: University of South Carolina Press, 2006.
Wellman, Manly Wade. *Fastest on the River*. New York: Henry Holt, 1957.

Newspapers

Louisville Courier-Journal
Memphis Appeal
New Orleans *Daily Picayune*
St. Louis Democrat
St. Louis Republican

Index

Álvarez de Piñeda, Alonso 19
barbers 118, 127
Barlow, Joel 54, 58, 61
Barlow, Ruth 54–55, 61
bars 127, 131
bartenders 127
Baton Rouge, La. 28, 41–43, 69, 76, 102
Bayou Sara, La. 28, 43
Beidenharm, J.A. 30
Ben Sherrod 128, 147–149
Berry, Tom 36, 39
Bofinger, John N. 134
Bonaparte, Napoleon 25, 52
British invasion 80–82
Brown, William Wells 117–118, 127
Buchanan, Thomas C. 118
Burnham, Mort 45, 180
Cabeza de Vaca, A.N. 19
Cairo, Ill. 16, 32, 160, 163, 173, 175–180, 189
calliope 110
Cannon, John W. 7–8, 10–18, 35–42, 44, 146–147, 167–168, 170–171, 174–177, 179–182, 185–186, 189–192, 194–195
Cannon, William 194–195
Cape Girardeau, Mo. 32, 179–180, 189
captains 129–137, 140, 144, 151
cargo 97–99
Cass, Lewis 23
Cass Lake 23
chambermaids 126
Chapman family 197
Chevalier, Michael 151
Cincinnati, Ohio 66–67, 70–71, 85, 87, 94, 150, 153, 159, 175, 189, 195
Cincinnati Gazette 189
Clayton, Frank 45, 180
Clayton, George 176, 181, 192
Clemens, Samuel L. 5, 28, 30, 101–102, 122–123, 144, 198
clerks 131, 138, 144

Clermont 64
Clermont, N.Y. 49, 64
Coca-Cola 30
Comet 78, 82, 84
construction improvements 95–96
cooks 124–126
crew, meals 119–121; racial and ethnic composition 118–119; racial violence 119; segregation 119–120; slaves 117–119, 126
cuisine 106
Daily Picayune 3–4, 45–47, 119, 151, 169, 175, 185
Daily States 196
deckhands 118, 121, 123–124, 137
description, steamboat 83, 96, 102–108
de Soto, Hernando 19–20
Devil's Country 179
Devol, George H. 46, 114–115
Dickens, Charles 90, 109
Donaldsonville, La. 28, 38
Eads, James B. 186
Eclipse 104
Effie Afton 156–158
engineers 131, 139–140, 144
Enterprise 78–83, 85–86
firemen 123–124
first mates 137–138, 144
Frank Pargoud 167–168, 176, 188
freight rates 96, 132
French, Augustus Byron 136
French, Callie Leach 136
French, Daniel 77–79, 83–84
Fulton, Robert 27, 52–65, 69, 75, 77–78, 82, 84, 89, 91
gambling 112–116
George Washington 87
Goldenrod 197
Gould, E.W. 1, 195, 197
Grampus 145–146
Grand Tower 181–182
Greene, Mary Becker (Ma) 136

207

Index

Helena, Ark. 31, 47, 168–170
Hopkins, Arthur E. 102
Hudson River 49, 58–59, 62, 89–90, 129
ice 155
Idlewild 175–176
Jackson, Andrew 80–81
Jameson, Jesse 176, 180
Jefferson, Thomas 25, 52
J.M. White 10, 191–192
Joliet, Louis 20–21
King, Enoch 176, 180
Lake Itasca 24
Lake Providence, La. 30
LaSalle, R.R. Cavelier 21–23
Leathers, Blanche Douglass 135–136
Leathers, Bowling 135, 196
Leathers, Thomas P. 8–10, 42–47, 135–136, 167–169, 174, 178–182, 187–192, 195–196
Lincoln, Abraham 157–158, 160
Livingston, Edward 80, 82, 85–87
Livingston, John 75
Livingston, Robert R. 25, 49, 51–52, 54–62, 64, 69, 75, 77, 80, 89
Louisiana 11, 38, 146
Louisville, Ky. 67–68, 71–72, 77, 82, 85, 104, 175
Marquette, Père Jacques 20–21
Marshall, Chief Justice John 89
meals, service 120–121
Memphis, Tenn. 31, 149, 160, 163, 170–174, 194
Merrick, George Byron 110, 112–113, 124–125, 127, 132–133, 137–138, 144
Miller, Mary 136
Monroe, James 25
music, on steamboats 109–110
names, for steamboats 134–135
Napoleon, Ark. 30–31, 169
Natchez 1, 2, 4, 6–10, 12–18, 35, 38, 41–47, 135, 167–182, 185–191, 193–196
Natchez, Miss. 28–29, 44, 68, 74–75, 77, 126, 128, 147, 194, 196
New Madrid, Mo. 31, 68, 74, 174
New Orleans 27, 69–77, 79
New Orleans, La. 3–11, 13–16, 18, 25, 27, 35–36, 38, 42–43, 45, 47, 65–66, 68–69, 75, 77–82, 84–87, 98–99, 101–102, 104, 109, 115–117, 126, 128–131, 145–147, 160, 176, 184, 188, 190, 193–194, 196
Newcomen engine 50–51
North River Steam Boat 64, 89

Olmsted, Frederick Law 103
Pakenham, Edward 81
passenger rates 97, 109, 132
Perkins, William 36, 39
Perrier, Jacques 57–58
pilots 121–122, 133, 140, 142–144
Pittsburgh, Pa. 65–67, 70, 77–80, 83, 94, 115, 129
Plaquemine, La. 40–41
population growth 99
porters 126–127
Quick, Herbert and Edward 6, 140, 142–143
rate wars 132
receipts, steamboat 97–98
Reelfoot Lake 74
Robert E. Lee 1, 2, 4, 6–8, 11, 13–18, 35–47, 115, 167–194
Roosevelt, Lydia Latrobe 65–76
Roosevelt, Nicholas J. 51, 56, 65–76, 79
roustabouts 120–121, 124, 168
St. Genevieve, Mo. 32, 182
St. Louis, Mo. 4, 6–7, 13, 15, 32–34, 74–79, 101–102, 104, 117, 129, 154, 161, 176–178, 180–181, 183, 185–192, 194–195
St. Louis Democrat 183, 185–186, 188
St. Louis Republican 3, 41–43, 88, 168, 174–175, 178–179, 181, 187
St. Mary's Market 18, 35, 47, 174
Schoolcraft, Henry R. 23–24
Scott, Alex 129
Shreve, Henry Miller 78–87, 147
slaves 109, 117–118
Smyth, A.W. 109, 117–118
snags 154
soundings 121–123
statistics, steamboat 93–97
Stevens, John 51, 65, 91
stewards 125–127, 140–142
Sultana 159–166, 173
Teche 145
Thompson Dean 171
Tobin, John W. 10, 168
Trollope, Frances 29
Vicksburg, Miss. 30, 46–47, 148, 160–161, 167
waiters 127
Warmoth, Henry C. 16, 35–38, 168
Washington 83–87, 147
water, drinking 106
Watt, James 50–51, 54, 60
Wiest, John 39–40, 176, 180–181

www.ingramcontent.com/pod-product-compliance
Ingram Content Group UK Ltd.
Pitfield, Milton Keynes, MK11 3LW, UK
UKHW042001140426
5217IPUK00015B/917